State-of-the-Art AI Driverless Self-Driving Cars

Practical Advances in Artificial Intelligence (AI) and Machine Learning

Dr. Lance B. Eliot, MBA, PhD

Disclaimer: This book is presented solely for educational and entertainment purposes. The author and publisher are not offering it as legal, accounting, or other professional services advice. The author and publisher make no representations or warranties of any kind and assume no liabilities of any kind with respect to the accuracy or completeness of the contents and specifically disclaim any implied warranties of merchantability or fitness of use for a particular purpose. Neither the author nor the publisher shall be held liable or responsible to any person or entity with respect to any loss or incidental or consequential damages caused, or alleged to have been caused, directly or indirectly, by the information or programs contained herein. Every company is different and the advice and strategies contained herein may not be suitable for your situation.

DEDICATION

To my incredible daughter, Lauren and my incredible son, Michael.

Forest fortuna adiuvat (from the Latin; good fortune favors the brave).

CONTENTS

Lance B. Eliot

ACKNOWLEDGMENTS

I have been the beneficiary of advice and counsel by many friends, colleagues, family, investors, and many others. I want to thank everyone that has aided me throughout my career. I write from the heart and the head, having experienced first-hand what it means to have others around you that support you during the good times and the tough times.

To Warren Bennis, one of my doctoral advisors and ultimately a colleague, I offer my deepest thanks and appreciation, especially for his calm and insightful wisdom and support.

To Mark Stevens and his generous efforts toward funding and supporting the USC Stevens Center for Innovation.

To Lloyd Greif and the USC Lloyd Greif Center for Entrepreneurial Studies for their ongoing encouragement of founders and entrepreneurs.

To Peter Drucker, William Wang, Aaron Levie, Peter Kim, Jon Kraft, Cindy Crawford, Jenny Ming, Steve Milligan, Chis Underwood, Frank Gehry, Buzz Aldrin, Steve Forbes, Bill Thompson, Dave Dillon, Alan Fuerstman, Larry Ellison, Jim Sinegal, John Sperling, Mark Stevenson, Anand Nallathambi, Thomas Barrack, Jr., and many other innovators and leaders that I have met and gained mightily from doing so.

Thanks to Ed Trainor, Kevin Anderson, James Hickey, Wendell Jones, Ken Harris, DuWayne Peterson, Mike Brown, Jim Thornton, Abhi Beniwal, Al Biland, John Nomura, Eliot Weinman, John Desmond, and many others for their unwavering support during my career.

And most of all thanks as always to Lauren and Michael, for their ongoing support and for having seen me writing and heard much of this material during the many months involved in writing it. To their patience and willingness to listen.

Lance B. Eliot

INTRODUCTION

This is a book that provides the newest innovations and the latest Artificial Intelligence (AI) advances about the emerging nature of AI-based autonomous self-driving driverless cars. Via recent advances in Artificial Intelligence (AI) and Machine Learning (ML), we are nearing the day when vehicles can control themselves and will not require and nor rely upon human intervention to perform their driving tasks (or, that <u>allow</u> for human intervention, but only *require* human intervention in very limited ways).

Similar to my other related books, which I describe in a moment and list the chapters in the Appendix A of this book, I am particularly focused on those advances that pertain to self-driving cars. The phrase "autonomous vehicles" is often used to refer to any kind of vehicle, whether it is ground-based or in the air or sea, and whether it is a cargo hauling trailer truck or a conventional passenger car. Though the aspects described in this book are certainly applicable to all kinds of autonomous vehicles, I am focused more so here on cars.

Indeed, I am especially known for my role in aiding the advancement of self-driving cars, serving currently as the Executive Director of the Cybernetic Self-Driving Cars Institute.. In addition to writing software, designing and developing systems and software for self-driving cars, I also speak and write quite a bit about the topic. This book is a collection of some of my more advanced essays. For those of you that might have seen my essays posted elsewhere, I have updated them and integrated them into this book as one handy cohesive package.

You might be interested in companion books that I have written that provide additional key innovations and fundamentals about self-driving cars. Those books are entitled **"Introduction to Driverless Self-Driving Cars," "Advances in AI and Autonomous Vehicles: Cybernetic Self-Driving Cars," "Self-Driving Cars: "The Mother of All AI Projects," "Innovation and Thought Leadership on Self-Driving Driverless Cars," "New Advances in AI Autonomous Driverless Self-Driving Cars,"** and **"Autonomous Vehicle Driverless Self-Driving Cars and**

Artificial Intelligence" and **"Transformative Artificial Intelligence Driverless Self-Driving Cars,"** and **"Disruptive Artificial Intelligence and Driverless Self-Driving Cars"** (they are all available via Amazon). See Appendix A of this herein book to see a listing of the chapters covered in those three books.

For the introduction here to this book, I am going to borrow my introduction from those companion books, since it does a good job of laying out the landscape of self-driving cars and my overall viewpoints on the topic. The remainder of the book is all new material that does not appear in the companion books.

INTRODUCTION TO SELF-DRIVING CARS

This is a book about self-driving cars. Someday in the future, we'll all have self-driving cars and this book will perhaps seem antiquated, but right now, we are at the forefront of the self-driving car wave. Daily news bombards us with flashes of new announcements by one car maker or another and leaves the impression that within the next few weeks or maybe months that the self-driving car will be here. A casual non-technical reader would assume from these news flashes that in fact we must be on the cusp of a true self-driving car.

Here's a real news flash: We are still quite a distance from having a true self-driving car. It is years to go before we get there.

Why is that? Because a true self-driving car is akin to a moonshot. In the same manner that getting us to the moon was an incredible feat, likewise can it be said for achieving a true self-driving car. Anybody that suggests or even brashly states that the true self-driving car is nearly here should be viewed with great skepticism. Indeed, you'll see that I often tend to use the word "hogwash" or "crock" when I assess much of the decidedly *fake news* about self-driving cars. Those of us on the inside know that what is often reported to the outside is malarkey. Few of the insiders are willing to say so. I have no such hesitation.

Indeed, I've been writing a popular blog post about self-driving cars and hitting hard on those that try to wave their hands and pretend that we are on the imminent verge of true self-driving cars. For many years, I've been known as the AI Insider. Besides writing about AI, I also develop AI software. I do what I describe. It also gives me insights into what others that are doing AI are really doing versus what it is said they are doing.

Many faithful readers had asked me to pull together my insightful short essays and put them into another book, which you are now holding in your hands.

For those of you that have been reading my essays over the years, this collection not only puts them together into one handy package, I also updated the essays and added new material. For those of you that are new to the topic of self-driving cars and AI, I hope you find these essays approachable and informative. I also tend to have a writing style with a bit of a voice, and so you'll see that I am times have a wry sense of humor and also like to poke at conformity.

As a former professor and founder of an AI research lab, I for many years wrote in the formal language of academic writing. I published in referred journals and served as an editor for several AI journals. This writing here is not of the nature, and I have adopted a different and more informal style for these essays. That being said, I also do mention from time-to-time more rigorous material on AI and encourage you all to dig into those deeper and more formal materials if so interested.

I am also an AI practitioner. This means that I write AI software for a living. Currently, I head-up the Cybernetics Self-Driving Car Institute, where we are developing AI software for self-driving cars. I am excited to also report that my son, also a software engineer, heads-up our Cybernetics Self-Driving Car Lab. What I have helped to start, and for which he is an integral part, ultimately he will carry long into the future after I have retired. My daughter, a marketing whiz, also is integral to our efforts as head of our Marketing group. She too will carry forward the legacy now being formulated.

For those of you that are reading this book and have a penchant for writing code, you might consider taking a look at the open source code available for self-driving cars. This is a handy place to start learning how to develop AI for self-driving cars. There are also many new educational courses spring forth.

There is a growing body of those wanting to learn about and develop self-driving cars, and a growing body of colleges, labs, and other avenues by which you can learn about self-driving cars.

This book will provide a foundation of aspects that I think will get you ready for those kinds of more advanced training opportunities. If you've already taken those classes, you'll likely find these essays especially interesting as they offer a perspective that I am betting few other instructors or faculty offered to you. These are challenging essays that ask you to think beyond the conventional about self-driving cars.

THE MOTHER OF ALL AI PROJECTS

In June 2017, Apple CEO Tim Cook came out and finally admitted that Apple has been working on a self-driving car. As you'll see in my essays, Apple was enmeshed in secrecy about their self-driving car efforts. We have only been able to read the tea leaves and guess at what Apple has been up to. The notion of an iCar has been floating for quite a while, and self-driving engineers and researchers have been signing tight-lipped Non-Disclosure Agreements (NDA's) to work on projects at Apple that were as shrouded in mystery as any military invasion plans might be.

Tim Cook said something that many others in the Artificial Intelligence (AI) field have been saying, namely, the creation of a self-driving car has got to be the mother of all AI projects. In other words, it is in fact a tremendous moonshot for AI. If a self-driving car can be crafted and the AI works as we hope, it means that we have made incredible strides with AI and that therefore it opens many other worlds of potential breakthrough accomplishments that AI can solve.

Is this hyperbole? Am I just trying to make AI seem like a miracle worker and so provide self-aggrandizing statements for those of us writing the AI software for self-driving cars? No, it is not hyperbole. Developing a true self-driving car is really, really, really hard to do. Let me take a moment to explain why. As a side note, I realize that the Apple CEO is known for at times uttering hyperbole, and he had previously said for example that the year 2012 was "the mother of all years," and he had said that the release of iOS 10 was "the mother of all releases" – all of which does suggest he likes to use the handy "mother of" expression. But, I assure you, in terms of true self-driving cars, he has hit the nail on the head. For sure.

When you think about a moonshot and how we got to the moon, there are some identifiable characteristics and those same aspects can be applied to creating a true self-driving car. You'll notice that I keep putting the word "true" in front of the self-driving car expression. I do so because as per my essay about the various levels of self-driving cars (see Chapter 3), there are some self-driving cars that are only somewhat of a self-driving car. The somewhat versions are ones that require a human driver to be ready to intervene. In my view, that's not a true self-driving car. A true self-driving car is one that requires no human driver intervention at all. It is a car that can entirely undertake via automation the driving task without any human driver needed. This is the essence of what is known as a Level 5 self-driving car. We are currently at the Level 2 and Level 3 mark, and not yet at Level 5.

Getting to the moon involved aspects such as having big stretch goals, incremental progress, experimentation, innovation, and so on. Let's review how this applied to the moonshot of the bygone era, and how it applies to the self-driving car moonshot of today.

Big Stretch Goal

Trying to take a human and deliver the human to the moon, and bring them back, safely, was an extremely large stretch goal at the time. No one knew whether it could be done. The technology wasn't available yet. The cost was huge. The determination would need to be fierce. Etc. To reach a Level 5 self-driving car is going to be the same. It is a big stretch goal. We can readily get to the Level 3, and we are able to see the Level 4 just up ahead, but a Level 5 is still an unknown as to if it is doable. It should eventually be doable and in the same way that we thought we'd eventually get to the moon, but when it will occur is a different story.

Incremental Progress

Getting to the moon did not happen overnight in one fell swoop. It took years and years of incremental progress to get there. Likewise for self-driving cars. Google has famously been striving to get to the Level 5, and pretty much been willing to forgo dealing with the intervening levels, but most of the other self-driving car makers are doing the incremental route. Let's get a good Level 2 and a somewhat Level 3 going. Then, let's improve the Level 3 and get a somewhat Level 4 going. Then, let's improve the Level 4 and finally arrive at a Level 5. This seems to be the prevalent way that we are going to achieve the true self-driving car.

Experimentation

You likely know that there were various experiments involved in perfecting the approach and technology to get to the moon. As per making incremental progress, we first tried to see if we could get a rocket to go into space and safety return, then put a monkey in there, then with a human, then we went all the way to the moon but didn't land, and finally we arrived at the mission that actually landed on the moon. Self-driving cars are the same way. We are doing simulations of self-driving cars. We do testing of self-driving cars on private land under controlled situations. We do testing of self-driving cars on public roadways, often having to meet regulatory requirements including for example having an engineer or equivalent in the car to take over the controls if needed. And so on. Experiments big and small are needed to figure out what works and what doesn't.

Innovation

There are already some advances in AI that are allowing us to progress toward self-driving cars. We are going to need even more advances. Innovation in all aspects of technology are going to be required to achieve a true self-driving car. By no means do we already have everything in-hand that we need to get there. Expect new inventions and new approaches, new algorithms, etc.

Setbacks

Most of the pundits are avoiding talking about potential setbacks in the progress toward self-driving cars. Getting to the moon involved many setbacks, some of which you never have heard of and were buried at the time so as to not dampen enthusiasm and funding for getting to the moon. A recurring theme in many of my included essays is that there are going to be setbacks as we try to arrive at a true self-driving car. Take a deep breath and be ready. I just hope the setbacks don't completely stop progress. I am sure that it will cause progress to alter in a manner that we've not yet seen in the self-driving car field. I liken the self-driving car of today to the excitement everyone had for Uber when it first got going. Today, we have a different view of Uber and with each passing day there are more regulations to the ride sharing business and more concerns raised. The darling child only stays a darling until finally that child acts up. It will happen the same with self-driving cars.

SELF-DRIVING CARS CHALLENGES

But what exactly makes things so hard to have a true self-driving car, you might be asking. You have seen cruise control for years and years. You've lately seen cars that can do parallel parking. You've seen YouTube videos of Tesla drivers that put their hands out the window as their car zooms along the highway, and seen to therefore be in a self-driving car. Aren't we just needing to put a few more sensors onto a car and then we'll have in-hand a true self-driving car? Nope.

Consider for a moment the nature of the driving task. We don't just let anyone at any age drive a car. Worldwide, most countries won't license a driver until the age of 18, though many do allow a learner's permit at the age of 15 or 16. Some suggest that a younger age would be physically too small to reach the controls of the car. Though this might be the case, we could easily adjust the controls to allow for younger aged and thus smaller stature. It's not their physical size that matters. It's their cognitive development that matters.

To drive a car, you need to be able to reason about the car, what the car can and cannot do. You need to know how to operate the car. You need to know about how other cars on the road drive. You need to know what is allowed in driving such as speed limits and driving within marked lanes. You need to be able to react to situations and be able to avoid getting into accidents. You need to ascertain when to hit your brakes, when to steer clear of a pedestrian, and how to keep from ramming that motorcyclist that just cut you off.

Many of us had taken courses on driving. We studied about driving and took driver training. We had to take a test and pass it to be able to drive. The point being that though most adults take the driving task for granted, and we often "mindlessly" drive our cars, there is a significant amount of cognitive effort that goes into driving a car. After a while, it becomes second nature. You don't especially think about how you drive, you just do it. But, if you watch a novice driver, say a teenager learning to drive, you suddenly realize that there is a lot more complexity to it than we seem to realize.

Furthermore, driving is a very serious task. I recall when my daughter and son first learned to drive. They are both very conscientious people. They wanted to make sure that whatever they did, they did well, and that they did not harm anyone. Every day, when you get into a car, it is probably around 4,000 pounds of hefty metal and plastics (about two tons), and it is a lethal weapon. Think about it. You drive down the street in an object that weighs two tons and with the engine it can accelerate and ram into anything you want to hit. The damage a car can inflict is very scary. Both my children were surprised that they were being given the right to maneuver this monster of a beast that could cause tremendous harm entirely by merely letting go of the steering wheel for a moment or taking your eyes off the road.

In fact, in the United States alone there are about 30,000 deaths per year by auto accidents, which is around 100 per day. Given that there are about 263 million cars in the United States, I am actually more amazed that the number of fatalities is not a lot higher. During my morning commute, I look at all the thousands of cars on the freeway around me, and I think that if all of them decided to go zombie and drive in a crazy maniac way, there would be many people dead. Somehow, incredibly, each day, most people drive relatively safely. To me, that's a miracle right there. Getting millions and millions of people to be safe and sane when behind the wheel of a two ton mobile object, it's a feat that we as a society should admire with pride.

So, hopefully you are in agreement that the driving task requires a great deal of cognition. You don't' need to be especially smart to drive a car, and we've done quite a bit to make car driving viable for even the average dolt. There isn't an IQ test that you need to take to drive a car. If you can read and write, and pass a test, you pretty much can legally drive a car. There are of course some that drive a car and are not legally permitted to do so, plus there are private areas such as farms where drivers are young, but for public roadways in the United States, you can be generally of average intelligence (or less) and be able to legally drive.

This though makes it seem like the cognitive effort must not be much. If the cognitive effort was truly hard, wouldn't we only have Einstein's that could drive a car? We have made sure to keep the driving task as simple as we can, by making the controls easy and relatively standardized, and by having roads that are relatively standardized, and so on. It is as though Disneyland has put their Autopia into the real-world, by us all as a society agreeing that roads will be a certain way, and we'll all abide by the various rules of driving.

A modest cognitive task by a human is still something that stymies AI. You certainly know that AI has been able to beat chess players and be good at other kinds of games. This type of narrow cognition is not what car driving is about. Car driving is much wider. It requires knowledge about the world, which a chess playing AI system does not need to know. The cognitive aspects of driving are on the one hand seemingly simple, but at the same time require layer upon layer of knowledge about cars, people, roads, rules, and a myriad of other "common sense" aspects. We don't have any AI systems today that have that same kind of breadth and depth of awareness and knowledge.

As revealed in my essays, the self-driving car of today is using trickery to do particular tasks. It is all very narrow in operation. Plus, it currently assumes that a human driver is ready to intervene. It is like a child that we have taught to stack blocks, but we are needed to be right there in case the child stacks them too high and they begin to fall over. AI of today is brittle, it is narrow, and it does not approach the cognitive abilities of humans. This is why the true self-driving car is somewhere out in the future.

Another aspect to the driving task is that it is not solely a mind exercise. You do need to use your senses to drive. You use your eyes a vision sensors to see the road ahead. You vision capability is like a streaming video, which your brain needs to continually analyze as you drive. Where is the road? Is there a pedestrian in the way? Is there another car ahead of you? Your senses are relying a flood of info to your brain. Self-driving cars are trying to do the same, by using cameras, radar, ultrasound, and lasers. This is an attempt at mimicking how humans have senses and sensory apparatus.

Thus, the driving task is mental and physical. You use your senses, you use your arms and legs to manipulate the controls of the car, and you use your brain to assess the sensory info and direct your limbs to act upon the controls of the car. This all happens instantly. If you've ever perhaps gotten something in your eye and only had one eye available to drive with, you suddenly realize how dependent upon vision you are. If you have a broken foot with a cast, you suddenly realize how hard it is to control the brake pedal and the accelerator. If you've taken medication and your brain is maybe sluggish, you suddenly realize how much mental strain is required to drive a car.

An AI system that plays chess only needs to be focused on playing chess. The physical aspects aren't important because usually a human moves the chess pieces or the chessboard is shown on an electronic display. Using AI for a more life-and-death task such as analyzing MRI images of patients, this again does not require physical capabilities and instead is done by examining images of bits.

Driving a car is a true life-and-death task. It is a use of AI that can easily and at any moment produce death. For those colleagues of mine that are developing this AI, as am I, we need to keep in mind the somber aspects of this. We are producing software that will have in its virtual hands the lives of the occupants of the car, and the lives of those in other nearby cars, and the lives of nearby pedestrians, etc. Chess is not usually a life-or-death matter.

Driving is all around us. Cars are everywhere. Most of today's AI applications involve only a small number of people. Or, they are behind the scenes and we as humans have other recourse if the AI messes up. AI that is driving a car at 80 miles per hour on a highway had better not mess up. The consequences are grave. Multiply this by the number of cars, if we could put magically self-driving into every car in the USA, we'd have AI running in the 263 million cars. That's a lot of AI spread around. This is AI on a massive scale that we are not doing today and that offers both promise and potential peril.

There are some that want AI for self-driving cars because they envision a world without any car accidents. They envision a world in which there is no car congestion and all cars cooperate with each other. These are wonderful utopian visions.

They are also very misleading. The adoption of self-driving cars is going to be incremental and not overnight. We cannot economically just junk all existing cars. Nor are we going to be able to affordably retrofit existing cars. It is more likely that self-driving cars will be built into new cars and that over many years of gradual replacement of existing cars that we'll see the mix of self-driving cars become substantial in the real-world.

In these essays, I have tried to offer technological insights without being overly technical in my description, and also blended the business, societal, and economic aspects too. Technologists need to consider the non-technological impacts of what they do. Non-technologists should be aware of what is being developed.

We all need to work together to collectively be prepared for the enormous disruption and transformative aspects of true self-driving cars. We all need to be involved in this mother of all AI projects.

WHAT THIS BOOK PROVIDES

What does this book provide to you? It introduces many of the key elements about self-driving cars and does so with an AI based perspective. I weave together technical and non-technical aspects, readily going from being concerned about the cognitive capabilities of the driving task and how the technology is embodying this into self-driving cars, and in the next breath I discuss the societal and economic aspects.

They are all intertwined because that's the way reality is. You cannot separate out the technology per se, and instead must consider it within the milieu of what is being invented and innovated, and do so with a mindset towards the contemporary mores and culture that shape what we are doing and what we hope to do.

WHY THIS BOOK

I wrote this book to try and bring to the public view many aspects about self-driving cars that nobody seems to be discussing.

For business leaders that are either involved in making self-driving cars or that are going to leverage self-driving cars, I hope that this book will enlighten you as to the risks involved and ways in which you should be strategizing about how to deal with those risks.

For entrepreneurs, startups and other businesses that want to enter into the self-driving car market that is emerging, I hope this book sparks your interest in doing so, and provides some sense of what might be prudent to pursue.

For researchers that study self-driving cars, I hope this book spurs your interest in the risks and safety issues of self-driving cars, and also nudges you toward conducting research on those aspects.

For students in computer science or related disciplines, I hope this book will provide you with interesting and new ideas and material, for which you might conduct research or provide some career direction insights for you.

For AI companies and high-tech companies pursuing self-driving cars, this book will hopefully broaden your view beyond just the mere coding and development needed to make self-driving cars.

For all readers, I hope that you will find the material in this book to be stimulating. Some of it will be repetitive of things you already know. But I am pretty sure that you'll also find various eureka moments whereby you'll discover a new technique or approach that you had not earlier thought of. I am also betting that there will be material that forces you to rethink some of your current practices.

I am not saying you will suddenly have an epiphany and change what you are doing. I do think though that you will reconsider or perhaps revisit what you are doing.

For anyone choosing to use this book for teaching purposes, please take a look at my suggestions for doing so, as described in the Appendix. I have found the material handy in courses that I have taught, and likewise other faculty have told me that they have found the material handy, in some cases as extended readings and in other instances as a core part of their course (depending on the nature of the class).

In my writing for this book, I have tried carefully to blend both the practitioner and the academic styles of writing. It is not as dense as is typical academic journal writing, but at the same time offers depth by going into the nuances and trade-offs of various practices.

The word "deep" is in vogue today, meaning getting deeply into a subject or topic, and so is the word "unpack" which means to tease out the underlying aspects of a subject or topic. I have sought to offer material that addresses an issue or topic by going relatively deeply into it and make sure that it is well unpacked.

Finally, in any book about AI, it is difficult to use our everyday words without having some of them be misinterpreted. Specifically, it is easy to anthropomorphize AI. When I say that an AI system "knows" something, I do not want you to construe that the AI system has sentience and "knows" in the same way that humans do. They aren't that way, as yet. I have tried to use quotes around such words from time-to-time to emphasize that the words I am using should not be misinterpreted to ascribe true human intelligence to the AI systems that we know of today. If I used quotes around all such words, the book would be very difficult to read, and so I am doing so judiciously. Please keep that in mind as you read the material, thanks.

Lance B. Eliot

COMPANION BOOKS

If you find this material of interest, you might want to also see my other books on self-driving cars, entitled:

1. **"Introduction to Driverless Self-Driving Cars"** by Dr. Lance Eliot

2. **"Innovation and Thought Leadership on Self-Driving Driverless Cars"** by Dr. Lance Eliot

3. **"Advances in AI and Autonomous Vehicles: Cybernetic Self-Driving Cars"** by Dr. Lance Eliot

4. *"Self-Driving Cars: The Mother of All AI Projects"* by Dr. Lance Eliot

5. **"New Advances in AI Autonomous Driverless Self-Driving Cars"** by Dr. Lance Eliot

6. **"Autonomous Vehicle Driverless Self-Driving Cars and Artificial Intelligence"** by Dr. Lance Eliot and Michael B. Eliot

7. **"Transformative Artificial Intelligence Driverless Self-Driving Cars"** by Dr. Lance Eliot

8. **"Disruptive Artificial Intelligence and Driverless Self-Driving Cars"** by Dr. Lance Eliot

All of the above books are available on Amazon and at other major global booksellers.

Lance B. Eliot

CHAPTER 1

ELIOT FRAMEWORK FOR AI SELF-DRIVING CARS

CHAPTER 1

ELIOT FRAMEWORK FOR AI SELF-DRIVING CARS

This chapter is a core foundational aspect for understanding AI self-driving cars and I have used this same chapter in several of my other books to introduce the reader to essential elements of this field. Once you've read this chapter, you'll be prepared to read the rest of the material since the foundational essence of the components of autonomous AI driverless self-driving cars will have been established for you.

———————

When I give presentations about self-driving cars and teach classes on the topic, I have found it helpful to provide a framework around which the various key elements of self-driving cars can be understood and organized (see diagram at the end of this chapter). The framework needs to be simple enough to convey the overarching elements, but at the same time not so simple that it belies the true complexity of self-driving cars. As such, I am going to describe the framework here and try to offer in a thousand words (or more!) what the framework diagram itself intends to portray.

The core elements on the diagram are numbered for ease of reference. The numbering does not suggest any kind of prioritization of the elements. Each element is crucial. Each element has a purpose, and otherwise would not be included in the framework. For some self-driving cars, a particular element might be more important or somehow distinguished in comparison to other self-driving cars.

You could even use the framework to rate a particular self-driving car, doing so by gauging how well it performs in each of the elements of the framework. I will describe each of the elements, one at a time. After doing so, I'll discuss aspects that illustrate how the elements interact and perform during the overall effort of a self-driving car.

At the Cybernetic Self-Driving Car Institute, we use the framework to keep track of what we are working on, and how we are developing software that fills in what is needed to achieve Level 5 self-driving cars.

D-01: Sensor Capture

Let's start with the one element that often gets the most attention in the press about self-driving cars, namely, the sensory devices for a self-driving car.

On the framework, the box labeled as D-01 indicates "Sensor Capture" and refers to the processes of the self-driving car that involve collecting data from the myriad of sensors that are used for a self-driving car. The types of devices typically involved are listed, such as the use of mono cameras, stereo cameras, LIDAR devices, radar systems, ultrasonic devices, GPS, IMU, and so on.

These devices are tasked with obtaining data about the status of the self-driving car and the world around it. Some of the devices are continually providing updates, while others of the devices await an indication by the self-driving car that the device is supposed to collect data. The data might be first transformed in some fashion by the device itself, or it might instead be fed directly into the sensor capture as raw data. At that point, it might be up to the sensor capture processes to do transformations on the data. This all varies depending upon the nature of the devices being used and how the devices were designed and developed.

D-02: Sensor Fusion

Imagine that your eyeballs receive visual images, your nose receives odors, your ears receive sounds, and in essence each of your distinct sensory devices is getting some form of input. The input befits the nature of the device. Likewise, for a self-driving car, the cameras provide visual images, the radar returns radar reflections, and so on.

Each device provides the data as befits what the device does.

At some point, using the analogy to humans, you need to merge together what your eyes see, what your nose smells, what your ears hear, and piece it all together into a larger sense of what the world is all about and what is happening around you. Sensor fusion is the action of taking the singular aspects from each of the devices and putting them together into a larger puzzle.

Sensor fusion is a tough task. There are some devices that might not be working at the time of the sensor capture. Or, there might some devices that are unable to report well what they have detected. Again, using a human analogy, suppose you are in a dark room and so your eyes cannot see much. At that point, you might need to rely more so on your ears and what you hear. The same is true for a self-driving car. If the cameras are obscured due to snow and sleet, it might be that the radar can provide a greater indication of what the external conditions consist of.

In the case of a self-driving car, there can be a plethora of such sensory devices. Each is reporting what it can. Each might have its difficulties. Each might have its limitations, such as how far ahead it can detect an object. All of these limitations need to be considered during the sensor fusion task.

D-03: Virtual World Model

For humans, we presumably keep in our minds a model of the world around us when we are driving a car. In your mind, you know that the car is going at say 60 miles per hour and that you are on a freeway. You have a model in your mind that your car is surrounded by other cars, and that there are lanes to the freeway. Your model is not only based on what you can see, hear, etc., but also what you know about the nature of the world. You know that at any moment that car ahead of you can smash on its brakes, or the car behind you can ram into your car, or that the truck in the next lane might swerve into your lane.

The AI of the self-driving car needs to have a virtual world model, which it then keeps updated with whatever it is receiving from the sensor fusion, which received its input from the sensor capture and the sensory devices.

D-04: System Action Plan

By having a virtual world model, the AI of the self-driving car is able to keep track of where the car is and what is happening around the car. In addition, the AI needs to determine what to do next. Should the self-driving car hit its brakes? Should the self-driving car stay in its lane or swerve into the lane to the left? Should the self-driving car accelerate or slow down?

A system action plan needs to be prepared by the AI of the self-driving car. The action plan specifies what actions should be taken. The actions need to pertain to the status of the virtual world model. Plus, the actions need to be realizable.

This realizability means that the AI cannot just assert that the self-driving car should suddenly sprout wings and fly. Instead, the AI must be bound by whatever the self-driving car can actually do, such as coming to a halt in a distance of X feet at a speed of Y miles per hour, rather than perhaps asserting that the self-driving car come to a halt in 0 feet as though it could instantaneously come to a stop while it is in motion.

D-05: Controls Activation

The system action plan is implemented by activating the controls of the car to act according to what the plan stipulates. This might mean that the accelerator control is commanded to increase the speed of the car. Or, the steering control is commanded to turn the steering wheel 30 degrees to the left or right.

One question arises as to whether or not the controls respond as they are commanded to do. In other words, suppose the AI has commanded the accelerator to increase, but for some reason it does not do so. Or, maybe it tries to do so, but the speed of the car does not increase. The controls activation feeds back into the virtual world model, and simultaneously the virtual world model is getting updated from the sensors, the sensor capture, and the sensor fusion. This allows the AI to ascertain what has taken place as a result of the controls being commanded to take some kind of action.

By the way, please keep in mind that though the diagram seems to have a linear progression to it, the reality is that these are all aspects of the self-driving car that are happening in parallel and simultaneously. The sensors are capturing data, meanwhile the sensor fusion is taking place, meanwhile the virtual model is being updated, meanwhile the system action plan is being formulated and reformulated, meanwhile the controls are being activated.

This is the same as a human being that is driving a car. They are eyeballing the road, meanwhile they are fusing in their mind the sights, sounds, etc., meanwhile their mind is updating their model of the world around them, meanwhile they are formulating an action plan of what to do, and meanwhile they are pushing their foot onto the pedals and steering the car. In the normal course of driving a car, you are doing all of these at once. I mention this so that when you look at the diagram, you will think of the boxes as processes that are all happening at the same time, and not as though only one happens and then the next.

They are shown diagrammatically in a simplistic manner to help comprehend what is taking place. You though should also realize that they are working in parallel and simultaneous with each other. This is a tough aspect in that the inter-element communications involve latency and other aspects that must be taken into account. There can be delays in one element updating and then sharing its latest status with other elements.

D-06: Automobile & CAN

Contemporary cars use various automotive electronics and a Controller Area Network (CAN) to serve as the components that underlie the driving aspects of a car. There are Electronic Control Units (ECU's) which control subsystems of the car, such as the engine, the brakes, the doors, the windows, and so on.

The elements D-01, D-02, D-03, D-04, D-05 are layered on top of the D-06, and must be aware of the nature of what the D-06 is able to do and not do.

D-07: In-Car Commands

Humans are going to be occupants in self-driving cars. In a Level 5 self-driving car, there must be some form of communication that takes place between the humans and the self-driving car. For example, I go into a self-driving car and tell it that I want to be driven over to Disneyland, and along the way I want to stop at In-and-Out Burger. The self-driving car now parses what I've said and tries to then establish a means to carry out my wishes.

In-car commands can happen at any time during a driving journey. Though my example was about an in-car command when I first got into my self-driving car, it could be that while the self-driving car is carrying out the journey that I change my mind. Perhaps after getting stuck in traffic, I tell the self-driving car to forget about getting the burgers and just head straight over to the theme park. The self-driving car needs to be alert to in-car commands throughout the journey.

D-08: VX2 Communications

We will ultimately have self-driving cars communicating with each other, doing so via V2V (Vehicle-to-Vehicle) communications. We will also have self-driving cars that communicate with the roadways and other aspects of the transportation infrastructure, doing so via V2I (Vehicle-to-Infrastructure).

The variety of ways in which a self-driving car will be communicating with other cars and infrastructure is being called V2X, whereby the letter X means whatever else we identify as something that a car should or would want to communicate with. The V2X communications will be taking place simultaneous with everything else on the diagram, and those other elements will need to incorporate whatever it gleans from those V2X communications.

D-09: Deep Learning

The use of Deep Learning permeates all other aspects of the self-driving car. The AI of the self-driving car will be using deep learning to do a better job at the systems action plan, and at the controls activation, and at the sensor fusion, and so on.

Currently, the use of artificial neural networks is the most prevalent form of deep learning. Based on large swaths of data, the neural networks attempt to "learn" from the data and therefore direct the efforts of the self-driving car accordingly.

D-10: Tactical AI

Tactical AI is the element of dealing with the moment-to-moment driving of the self-driving car. Is the self-driving car staying in its lane of the freeway? Is the car responding appropriately to the controls commands? Are the sensory devices working?

For human drivers, the tactical equivalent can be seen when you watch a novice driver such as a teenager that is first driving. They are focused on the mechanics of the driving task, keeping their eye on the road while also trying to properly control the car.

D-11: Strategic AI

The Strategic AI aspects of a self-driving car are dealing with the larger picture of what the self-driving car is trying to do. If I had asked that the self-driving car take me to Disneyland, there is an overall journey map that needs to be kept and maintained.

There is an interaction between the Strategic AI and the Tactical AI. The Strategic AI is wanting to keep on the mission of the driving, while the Tactical AI is focused on the particulars underway in the driving effort. If the Tactical AI seems to wander away from the overarching mission, the Strategic AI wants to see why and get things back on track. If the Tactical AI realizes that there is something amiss on the self-driving car, it needs to alert the Strategic AI accordingly and have an adjustment to the overarching mission that is underway.

D-12: Self-Aware AI

Very few of the self-driving cars being developed are including a Self-Aware AI element, which we at the Cybernetic Self-Driving Car Institute believe is crucial to Level 5 self-driving cars.

The Self-Aware AI element is intended to watch over itself, in the sense that the AI is making sure that the AI is working as intended. Suppose you had a human driving a car, and they were starting to drive erratically. Hopefully, their own self-awareness would make them realize they themselves are driving poorly, such as perhaps starting to fall asleep after having been driving for hours on end. If you had a passenger in the car, they might be able to alert the driver if the driver is starting to do something amiss. This is exactly what the Self-Aware

AI element tries to do, it becomes the overseer of the AI, and tries to detect when the AI has become faulty or confused, and then find ways to overcome the issue.

D-13: Economic

The economic aspects of a self-driving car are not per se a technology aspect of a self-driving car, but the economics do indeed impact the nature of a self-driving car. For example, the cost of outfitting a self-driving car with every kind of possible sensory device is prohibitive, and so choices need to be made about which devices are used. And, for those sensory devices chosen, whether they would have a full set of features or a more limited set of features.

We are going to have self-driving cars that are at the low-end of a consumer cost point, and others at the high-end of a consumer cost point. You cannot expect that the self-driving car at the low-end is going to be as robust as the one at the high-end. I realize that many of the self-driving car pundits are acting as though all self-driving cars will be the same, but they won't be. Just like anything else, we are going to have self-driving cars that have a range of capabilities. Some will be better than others. Some will be safer than others. This is the way of the real-world, and so we need to be thinking about the economics aspects when considering the nature of self-driving cars.

D-14: Societal

This component encompasses the societal aspects of AI which also impacts the technology of self-driving cars. For example, the famous Trolley Problem involves what choices should a self-driving car make when faced with life-and-death matters. If the self-driving car is about to either hit a child standing in the roadway, or instead ram into a tree at the side of the road and possibly kill the humans in the self-driving car, which choice should be made?

We need to keep in mind the societal aspects will underlie the AI of the self-driving car. Whether we are aware of it explicitly or not, the AI will have embedded into it various societal assumptions.

D-15: Innovation

I included the notion of innovation into the framework because we can anticipate that whatever a self-driving car consists of, it will continue to be innovated over time. The self-driving cars coming out in the next several years will undoubtedly be different and less innovative than the versions that come out in ten years hence, and so on.

Framework Overall

For those of you that want to learn about self-driving cars, you can potentially pick a particular element and become specialized in that aspect. Some engineers are focusing on the sensory devices. Some engineers focus on the controls activation. And so on. There are specialties in each of the elements.

Researchers are likewise specializing in various aspects. For example, there are researchers that are using Deep Learning to see how best it can be used for sensor fusion. There are other researchers that are using Deep Learning to derive good System Action Plans. Some are studying how to develop AI for the Strategic aspects of the driving task, while others are focused on the Tactical aspects.

A well-prepared all-around software developer that is involved in self-driving cars should be familiar with all of the elements, at least to the degree that they know what each element does. This is important since whatever piece of the pie that the software developer works on, they need to be knowledgeable about what the other elements are doing.

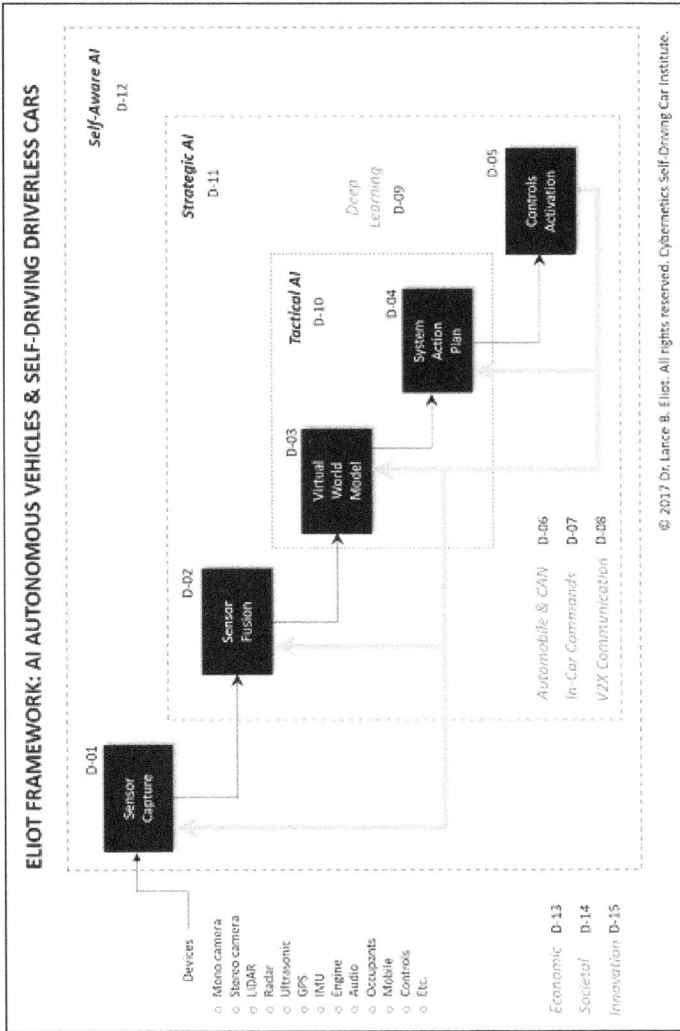

ELIOT FRAMEWORK: AI AUTONOMOUS VEHICLES & SELF-DRIVING DRIVERLESS CARS

Self-Aware AI
D-12

Strategic AI
D-11

Deep Learning
D-09

Tactical AI
D-10

Controls Activation
D-05

System Action Plan
D-04

Virtual World Model
D-03

Sensor Fusion
D-02

Sensor Capture
D-01

Devices
o Mono camera
o Stereo camera
o LIDAR
o Radar
o Ultrasonic
o GPS
o IMU
o Engine
o Audio
o Occupants
o Mobile
o Controls
o Etc.

Automobile & CAN   D-06
In-Car Commands   D-07
V2X Communication   D-08

Economic   D-13
Societal   D-14
Innovation D-15

CHAPTER 2

VERSIONING AND SELF-DRIVING CARS

Lance B. Eliot

CHAPTER 2

VERSIONING AND SELF-DRIVING CARS

Quick, tell me which version of Microsoft Windows you are running on your PC.

Is it Windows 10?

Or, maybe Window 8, Windows 7, Vista, XP, 2000, 98, NT, 95, etc. There have been numerous versions of Microsoft Windows that have been released over the years since its inception. You might have resisted upgrading at some point and still be on an older version of Windows, meanwhile the person seated next to you might be on the most recent version. Within versions, there are also releases. For Windows 10, you might be on release 1507 (code named Threshold 1), or 1511 (Threshold 2), or 1607 (Redstone 1), and so on.

Does it make a difference as to which version and which release of that version that you are running? Absolutely it makes a difference. Each version and each release has its own set of features and functionality. There are things you probably like about any one specific version and its releases, and things that you likely dislike. In spite of your own likes or dislikes, you pretty much get the whole kit and caboodle with a particular version/release and so you have to live with its good parts and its bad parts. Take it or leave it, that's the motto.

There are also bound to be bugs or errors in whichever version/release you are using. Some of those bugs or errors are known and published as being known. Some of those bugs and errors have patches or fixes that you can put in place to overcome the bug or error. There might be some bugs that aren't yet known and so they lurk within the system, waiting until possibly a bad moment to arise. Sometimes a version/release gets so riddled with bugs and errors, and has so many irritating "features" that you pine away waiting for the next version, in hopes that maybe you can dump the old one and adopt the new one.

Once again, though, the new one is ultimately going to have drawbacks in capabilities, along with both known and unknown bugs. You aren't going to wake-up and suddenly find that there's a version that does exactly what you want, in the way you want it, and that is totally bug free. Not going to happen.

What does this have to do with AI self-driving cars?

At the Cybernetic Self-Driving Car Institute, we are working on AI software that will be able to ascertain behaviors of AI self-driving cars as based on their versioning.

Allow me a moment to explain.

When AI self-driving cars begin to truly populate our traffic mix, they will not be homogeneous. By this I mean that not all AI self-driving cars will be the same. Some people mistakenly think that all AI self-driving cars will be identical. This false belief is predicated on the notion that every self-driving car will have the same AI components and software. You instead need to think about these AI systems in the same manner as you think about Microsoft Windows, namely there will be lots of different versions, many of which will be active at any given point in time.

Take a look at Figure 1.

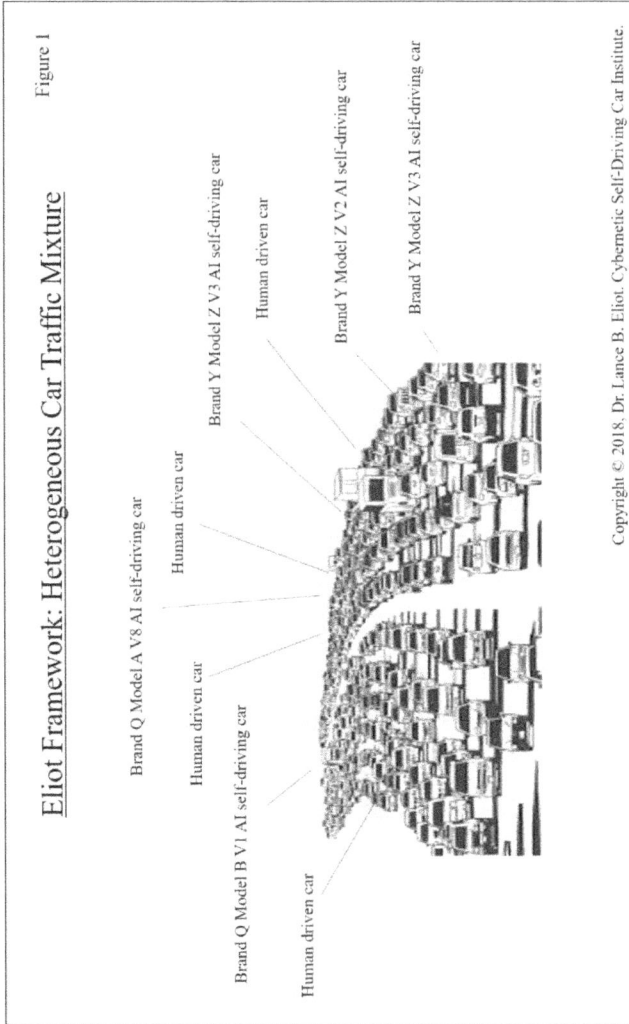

As shown, there are a number of cars in traffic. Let's assume that some of those cars are conventional cars and are being driven by humans. There are also some cars that are AI self-driving cars, of which, some are true Level 5 self-driving cars and others are less than Level 5. A true Level 5 self-driving car is one that can be driven entirely by the AI without human intervention and does not need any human driver for the undertaking of the driving task.

Let's assume we have an AI self-driving car that is Brand Q Model A and Version 8, and another AI self-driving car that is Brand Q Model B Version 1, and there's also a Brand Y Model Z Version 3, and a Brand Y Model Z Version 2. You can imagine that say the Brand Q is perhaps a Ford self-driving car, and the Brand Z is say a Toyota self-driving car.

The auto makers are going to have various brands of their self-driving cars and various models. Just like they do today for conventional cars. You might buy an AI self-driving car that is the latest version, while a friend of yours had bought one a few years earlier and has an earlier model. Even within the models of the AI self-driving cars, the AI software is going to be in different versions.

It's akin to having a room full of PC's that are running Microsoft Windows. Some are running version 10, some are running version 10 release 1507 and others are running version 10 release 1511. Other PC's in the room are running say version 8. And so on. The mix of AI self-driving cars in traffic will be just like this. Also, some of those PC's have the latest processors and other components, while some of those PC's have older processors and lack components that a more modern PC has. Some AI self-driving cars will have a radar and cameras that are of a particular type and model, while other AI self-driving cars will have different brands of radars and different brands of cameras. And so on.

I hope you are now past the idea of homogenous AI self-driving cars, and if so, you might be wondering what does it matter that on our roads we're going to have heterogeneous AI self-driving cars?

It matters for the same reason that knowing what version of Windows you are running on your PC, namely, the PC does different things, has differing functions and features, and contains different kinds of bugs and errors, some known and some unknown. The same is true for AI self-driving cars. The AI self-driving car that is Brand Q Model A that is running version 8, it might be known for being able to take turns very well, but it's not so good at handling roundabouts. The Brand Q Model B is a superior AI self-driving car to the Brand Q Model A in that it has newly added LIDAR (a sensor device that uses light and radar), which the Model B lacks. As such, the Model B can do a better job of detecting pedestrians and also spotting other cars further ahead than does the Model A.

I suppose you could think of this as having different models of conventional cars and having different kinds of drivers driving those cars. There are some human drivers that drive in a speedy fashion, and other human drivers that go slowly. There are some human drivers that seem to be able to see far off in the distance, and others that can barely see a few car lengths ahead. The capabilities of the AI self-driving cars will similarly vary.

For the AI pundits you might immediately object and say that with OTA (Over The Air) capabilities that we will be able to push new versions remotely to the AI self-driving cars. I agree that's going to be the case. But, it does not change the fact that the different auto makers will have different AI self-driving cars, and that besides having different physical car models with differing sensors on them, the AI systems running those models will differ.

That being said, if I am driving a Brand Y Model Z Version 3, and you are driving a Brand Y Model Z Version 2, it is perhaps likely that the Version 2 maybe has fallen behind doing its OTA and that it is pending a remote update so that it will be at Version 3. The OTA updates are not all going to happen at the same time to all of the AI self-driving cars in a particular brand and model. Perhaps I had done my OTA update to my beloved Brand Y Model Z in the morning before heading to work, but meanwhile your Brand Y Model Z was continually driving around and has not yet been put into a resting state to get its OTA.

It would be prudent and some say essential that the AI of a self-driving car be able to predict the behavior of other cars, both human driven cars and self-driving cars.

As shown in Figure 2, the prediction of other AI self-driving cars can be based on three methods: (1) looks like, and/or (2) by what it does (its behaviors), and/or (3) by V2V (vehicle to vehicle communications).

Let's consider each of these three methods.

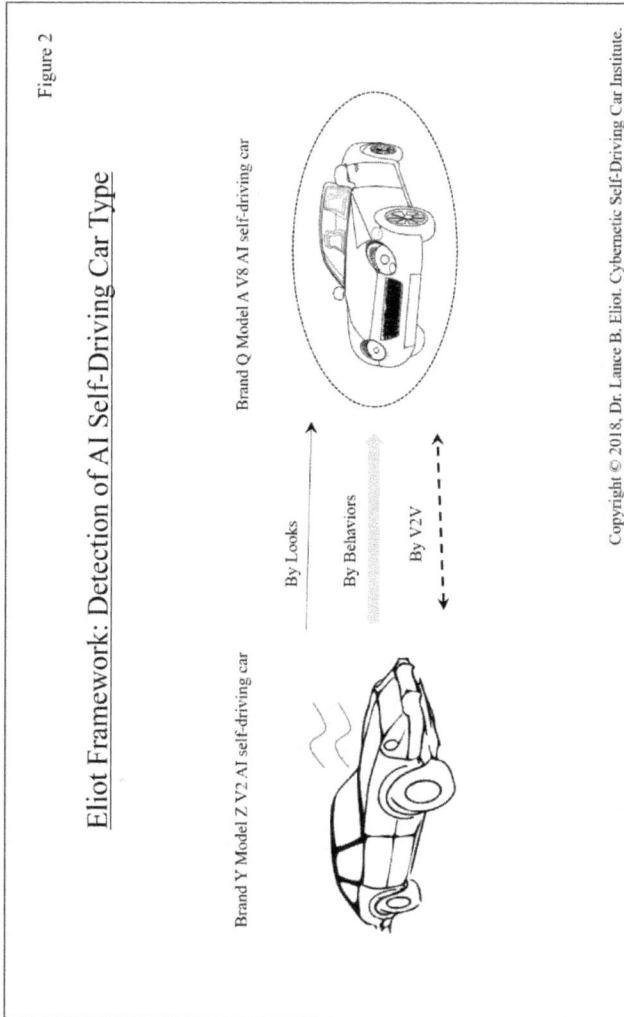

Figure 2

Eliot Framework: Detection of AI Self-Driving Car Type

Brand Q Model A V8 AI self-driving car

Brand Y Model Z V2 AI self-driving car

By Looks

By Behaviors

By V2V

First, when you see a Ford Mustang driving down the street, you instantly recognize the car due to its distinctive shape and styling. As such, for the AI self-driving cars, the auto makers are each taking their own approach to what their AI self-driving car will look like. Via shape and style alone, you can quickly gauge what auto maker made the car, what brand and model it is.

We do this via the sensors of our AI self-driving car, in that it uses the cameras to detect other cars and does an image analysis to identify what the other car is. Thus, the AI will be able to gauge readily what kinds of capabilities another AI self-driving car has, by typifying it via visual matching and then having a database of what the capabilities and limitations of that self-driving car is. This then allows the AI to predict what kinds of actions or moves that other AI self-driving car might make while on the road.

If the particular version of the AI self-driving car is not evident by visual inspection alone, another approach would be to observe its behavior. Suppose it is known that the Brand Y Model Z Version 2 often makes long stops at a stop sign, and it also is known for weaving far around a bicyclist, often going into another lane to try and get as far from a pedestrian as it can. Meanwhile, Version 3 has improvements that make the AI do a full stop at stop sign but not linger needlessly, and also that it does a better job of detecting the distance to a bicyclist and so it doesn't have to move over into another lane when things get tight.

The AI of our self-driving car observes the behaviors of other AI self-driving cars. And, based on the database of the limits and capabilities by model/brand/release, it is able to guess which release the other self-driving car is likely on. As such, the AI then also can better predict what the other AI self-driving car will do in various situations and circumstances.

Some AI pundits would say that there's no need to go to this much trouble about figuring out what the other AI self-driving car is and what it will do. They say that you should just ask it. With V2V, AI self-driving cars will be able to communicate directly with other AI self-driving cars. In that case, there's presumably no guessing needed. Instead, the AI of one car just asks the AI of the other car what kind of AI self-driving car it is. Furthermore, when any roadway action occurs, such as if there is a bicyclist up ahead, the AI can ask the other car what it is going to do once it gets near to the bicyclist.

Yes, there's no doubt that having V2V will allow for this kind of communication and coordination. Getting to that point though is somewhat unknown as yet. The industry is still working on the protocols for V2V. Also, the question arises as to how much V2V volume do we want and will we allow. Imagine if all AI self-driving cars are continually bombarding each other with requests about what they are doing. The computational effort to be responding to all these requests is going to chew-up both processing time and bandwidth.

Those that are strong proponents of V2V are also likely to make the assumption that V2V will always be available. It might not be. A particular self-driving car might not yet have V2V. Or, maybe a particular self-driving car has disabled some aspects of V2V. Or, maybe the communication link between two cars is not working well and so in spite of the two wanting to do V2V, they are stymied in doing so. The hope that V2V will be universal, will be always on, will never be disrupted, and will always work in all situations, I'd say it's a bit of false hope for now and that it would be more realistic to assume that V2V will intermittently be available.

If the V2V is intermittently available, we'd then have the other two approaches, the looks of the self-driving car and the behavior of the self-driving car. Thus, all three approaches can come together to try and predict what another AI self-driving car is doing and possibly going to do.

One other aspect about the heterogeneous nature of the AI self-driving cars will be how features will come and go, potentially. You might remember that Windows 7 had lots of nifty gadgets that allowed you to use a calculator or find out the weather status. Those were dropped in Windows 10. In a Darwinian process, some features are kept and others are dropped. The same will be true for AI self-driving cars.

Auto makers will try to differentiate their AI self-driving car over a competitor by claiming that their self-driving car does things that the other ones do not do. Prefer a really smooth ride in your AI self-driving car, the Brand X self-driving car has AI that is able to reduce reactions to bumps and potholes, and no other AI self-driving car has that same feature. It will be interesting to see how the features come and go. Would competitors perceive this smooth ride AI capability as essential, and therefore include it into a future version of their AI self-driving car.

There's also the bugs and errors aspects of AI self-driving cars. Suppose the Brand Y Model Z is known for being buggy. To-date, there have been several bugs found. Fortunately, those bugs were fixed and then pushed into the self-driving car via OTA. But, often where there are a few, there are more. It could be that the Brand Y Model Z has numerous other bugs or errors that just haven't been discovered. Or, maybe they are somewhat known, such as the Brand Y Model Z occasionally opts to slow down and speed-up, but no one has yet figured out why it does this.

Other AI self-driving cars can be on the watch for and wary of the behavior of the other AI self-driving cars.

They can do this by trying to gauge what the nature of the other AI self-driving car is. It's the same thing that we humans do. I am sure you've watched other cars around you and mentally thought that there's a car that might make a sudden turn, there's a car that will probably get in your way soon, and so on. True AI self-driving cars cannot just be driving along and acting as though there are other cars and getting caught off-guard by what those cars do. Instead, by paying attention to versioning of AI self-driving cars, every AI self-driving car can be preparing for the actions of their fellow AI self-driving cars.

CHAPTER 3

TOWING
AND
SELF-DRIVING CARS

CHAPTER 3

TOWING
AND SELF-DRIVING CARS

Towing something behind your car is not as easy as it might seem.

When I was in my teens, a good friend had a boat that we used to take with us when his family went on camping outings to the local lake. His father would take charge of making sure that the boat hitch was properly connected to the car, and his father also preferred to do all the driving since he told us that towing something requires special attention and special skills. We were pretty much fine with his doing this, and we figured it made life easier for us anyway. All we had to do was go along for the ride, and then enjoy doing some exciting water skiing and some quiet-time fishing once the boat magically got into the water at the lake.

I became somewhat painfully aware that there was a skill to doing towing when I later on got into college in Southern California and decided to help a friend tow his belongings up to the Bay Area where he was going to go to college. We went to the local U-Haul and rented one of those ubiquitous storage hitches. The attendant offered to help us connect it up, but we were too proud or too ignorant to realize that we should have welcomed his assistance. Thinking that two bright college aiming "adults" would be able to figure out what to do, we opted to do everything ourselves.

When we got to the grapevine, which is a big-rig filled, steep mountain passage that commonly is used to get from Los Angeles up to San Francisco, darned if we realized that the storage compartment was so filled with heavy items that the car could barely handle the strain of getting up the hills. We also heard a kind of grinding noise, and though loath to stop the car at this juncture on a steep hill, we pulled over and found out that we had not properly connected the hitch.

After getting it re-hitched, which was an unnerving chore on the busy highway, we were trying to now restart the towing from a complete standstill. The car began to overheat. We turned off the air conditioning in hopes that it would save the car some energy to use toward pulling the monstrosity that was connected to us. Meanwhile, the winds were whipping through the mountain passage and the storage compartment began to whipsaw. The fishtailing impact on the car was that the car itself began to become uncontrollable.

I don't know whether to look back upon this occasion as a fond memory, or consider it one of the most dangerous, and likely stupid things to have done. Should have been better prepared. Should have made sure the hitch was set right before we began the journey. Should have realized the wind would impact us. Should have made sure the car could handle the load. Should have known or practiced how to drive with a towing load. Should have done a ton of things. We instead used the simplistic bravado that's characteristic of the young and eager (we did eventually get to our destination, alive and well).

Let's then consider what should happen when you are going to tow something with your car.

First, there's the essential pre-towing preparations. Determine the towing capacity of your car, which can usually be found in the car owner's manual. Find out if your car can handle potentially shifting into lower gears. Ensure that the car's curb weight, which is when the car is empty and has a full tank of gas, and when added with the weight of any passengers and cargo, and add the tongue weight of the trailer, all of that should not be more than the gross vehicle weight rating (considered the total weight bearing down on the car's tires).

If needed, you might want to get your tires checked or replaced. You might want to get your brake pads checked or replaced. Make sure the oil is good. The coolant is good. Overall, will your car be able to handle the strain. Also, think about the path itself, such as if there are steep sloping roads and what the duration of the trip will be. Maybe your car can last for an hour, but will it be able to cope with eight hours of driving time?

Second, consider what to do about the pre-towing hitching. You need the right kind of trailer hitch for your car. There is the ball type of hitch. There's the fifth wheel type of hitch. There is the gooseneck hitch. And so on. Investigate how the hitch works. What kind of connection is required. How is the connection secured. Know your hitch like the back of your hand. When you then put the hitch on, double-check it. See that the trailer or whatever is being towed is properly connected and secured. Do you have safety chains? Do you have a lock?

Third, assuming that you are doing a trip like the one of taking items on a journey, you'll need to drive the car with the now hitched trailer vehicle to wherever you need to load it. We drove to my friend's house and realized that the car plus contraption would not fit in his driveway, and there wasn't any available street parking, so we double parked in the middle of the street. We thought we'd get away with this. Given our luck, it was not going to be so easy. About halfway of loading up the storage compartment, a police car drives up to us, tells us we have to immediately move. We begged that we just needed another twenty minutes to finish the chore. No doing, they said, move it or get busted.

Fourth, once the hitched vehicle is ready to roll, you are now able to proceed on your journey. You can't though drive like you normally do. There are aspects that change when you are towing something. I'll discuss this more in a moment.

Fifth, once you arrive at your destination, you'll likely have another parking chore to be undertaken and an unloading task too. When we used to go to the lake with the boat, the car had to back down a steep ramp, gradually lowering the hitch into the water. Once the water was floating the boat, we undid the boat from the hitch and gently pushed the boat into the water fully. The driver of the car was then signaled to drive up the ramp and go park the car someplace.

Sixth, eventually, once the journey has been completed, you'll likely want to unhitch the trailer and put it someplace, or turn it back into a rental facility. You need to be able to properly unhitch it. Properly store it. And if it is yours, likely do some maintenance so it will be in shape for the next use.

In short, we have these key major steps for towing:

- Pre-Towing preparations
- Pre-Towing hitching
- Towing journey start
- Towing journey mainstay
- Towing journey end
- Unhitch

What does this have to do with AI self-driving cars?

At the Cybernetic Self-Driving Car Institute, we are developing AI systems for self-driving cars, including having the AI be savvy enough to handle doing towing.

Towing is considered an edge problem by most auto makers and tech firms. An edge problem is one that is not at the core of something. The core right now for AI self-driving cars is to be able have a self-driving car that can drive in a normal fashion, being able to stay within its lanes, make lane changes, stop at stop signs, and so on. Handling a tow is not considered high priority.

Indeed, many AI developers tell me that they consider towing as already solved, since in their book if a car can drive normally, it can tow whatever it needs to tow. There isn't anything special about towing, they claim. Why make something out of nothing. No big deal. Just let the AI drive the car, tow or no tow.

We don't see things that way. Our view is that towing is a special case of car driving. When I was a naïve teenager, I certainly thought that towing was nothing unusual. I assumed the same falsehoods as these AI developers, namely you put the car in drive and away you go. As I later found out, there are differences. Important differences.

Here's what we have been developing as the AI "towing mode" capability:

- Self-driving car needs to go at a slower speed than normal, being extra cautious

- Stay in the slow lanes as much as possible, don't get into the fast lanes unless absolutely needed

- When lane changing, give extra distance and be watchful of other cars that might cut in

- Recalibrate stopping distances since the self-driving car can't stop as quickly as it could without the tow

- Recalculate stopping times as to how long it will take to bring the self-driving car to a halt

- Recalibrate starting times as to how quickly the self-driving car can reach certain speeds

- Be aware that starting from a stopped position will be longer and a strain on the self-driving car

- Recalibrate the total length of the self-driving car in terms of adding the length of the hitched item

- Try to detect fishtailing

- Drive to try and prevent fishtailing

- Be interactive with the occupants of the car for added insights about the towing

- Etc.

I won't go through each of the above aspects, but can provide you a glimpse at one of the elements, namely the notion of being watchful for other cars that might cut into a roadway opening when the self-driving car is trying to change lanes.

I'm betting that you've seen circumstances wherein a lengthy truck is trying to make a lane change, and some jerk car decides to zip right next to the truck and not allow the truck to readily make the lane change. The perspective of the car driver is apparently that if there is open roadway space, take it. Doesn't matter that the truck is signaling to get over. Sometimes it doesn't even matter if the truck has already started to make the maneuver and it is dangerous such that you might nearly strike the truck by zipping past it. There's many truck drivers that can attest to the lack of civility of car drivers (humans).

I mention this because a self-driving car with a hitched item is kind of the same thing as having a lengthy truck. The total length of the car plus the hitched item now makes lane changes unwieldly. More than unwieldly, it can be dangerous. The AI needs to realize that the methods of making lane changes that it has when the self-driving car is without a hitch are not quite the same with a hitched item.

This brings up a very important aspect about the self-driving car. When a self-driving car has a hitched item, the question arises as to how the sensors at the rear of the self-driving car will cope with the hitched item.

For example, the odds are that the cameras at the rear, which are there to detect what's behind the self-driving car, will only see the hitched item. No longer will those cameras provide an accurate depiction of what's really behind the self-driving car. Its view is blocked. The same can happen to the radar that's at the rear of the car. The same can be true of the sonar. Even the LIDAR, a combination of light and radar, will undoubtedly be partially blocked by whatever the hitched item is. Of course, if the hitched item has a lower profile, it's possible to see somewhat over it by some of the sensory devices. If it's tall and wide, the odds are that the sensors will all be pretty much blocked.

How can you make a lane change if those sensors cannot see what's behind the combined self-driving car and hitched item? Going on blind faith is obviously dangerous. There are some that would say that the self-driving car should not even be allowed to proceed if its rear sensors are blocked in this manner. Maybe there should even be a law that makes it illegal to tow when the self-driving car's rear sensory capability is degraded or unable to adequately perform.

This raises another question. How would the AI self-driving car even know that it is towing something? The AI isn't a person. It isn't going to just know that towing is happening. Now, it could certainly realize that the back sensors are being blocked, and so it could then opt to refuse to proceed. Or, maybe it would tell the owner or occupants that something is amiss at the rear end of the self-driving car.

Another aspect involves the owner or occupants telling outright to the AI that they are wanting to do a tow. The AI can then prepare itself for a towing mode. For a true self-driving car, which I consider a Level 5, which is a self-driving car that is supposed to be driven entirely by the AI and not require any human driver, we're likely going to be interacting verbally with the AI, using Natural Language Processing (NLP). Think of talking to the car akin to talking to Alexa or Siri. In this manner, you could have a dialogue with the AI, telling it that you are going to be towing. The AI if savvy enough would even ask questions about the nature of the towing.

This would allow the AI to figure out what paths for the journey might be best, such as avoiding steep hills, if feasible. The AI would also then switch into the mode of undertaking the driving elements I've mentioned earlier. It will go at a slower speed than normal. It will realize that stopping is going to be harder and take longer to do. And so on.

One way to deal with the rear sensors being blocked would be to have some kind of extra set of sensors that could be placed onto or at the end of the hitched item. I am anticipating we'll see towing kits for AI self-driving cars that provide this extra capability. It will be a little tricky because the question of how the extra sensors tie into the rest of the AI self-driving car needs to be determined, and also what the trustworthiness will be of those extra sensors. Suppose the human that attaches the sensors to the back of the hitched item does a lousy job and those sensors are now misreporting data to the AI system.

We'll also be including the V2V (vehicle to vehicle communications) aspects. The AI self-driving car will likely want to communicate electronically with other nearby self-driving cars to let those self-driving cars know that it is towing something and might need extra room. They can then electronically coordinate lane changes. This will hopefully do away with the lack of civility of human drivers. But, for quite a long time we're going to have a mix of human drivers with our AI self-driving cars, do don't be holding your breath that we're going to have all and only AI self-driving cars on the roadways anytime soon.

The AI needs to also be able to deal with situations such as the case of the hitch going awry, which I mentioned happened while we were on the grapevine. The AI might be able to detect that something is wrong via the sensors at the rear of the AI self-driving car, and then go into a mode of perhaps bringing the towing to a safe stop. Or, the occupants might alert the AI, via spoken commands, and ask the AI to safely pull over for an inspection.

We're gradually seeing conventional cars have things like "smart brakes" and "smart tires" that can report their status. This would be tied into the AI of the self-driving car. During towing, the AI should be monitoring the status of the brakes and the statue of the tires, using it to adjust how the driving is coming along. And, be able to hopefully predict that something might go awry, but do so before it actually occurs.

The first set of steps about preparing for a tow are still going to be in the hands of humans for the foreseeable future. It's still up to the human to make sure that their car is rightfully established for doing towing. The human still will need to put the hitch on. For now, only once the journey itself starts, the AI of the self-driving car mainly comes to play. I suppose we might have robots that can do this for us, but I'll bet that we'll have AI self-driving cars sooner than we have robots that will do so.

Towing with an AI self-driving car is something that people are going to assume they can do. Imagine that you bought this expensive AI self-driving car, and you later discover that you can't tow anything with it. The early adopters will probably accept this idea, and say that you can't expect the world of an AI self-driving car. As AI self-driving cars become more pervasive, I am sure that people will start to grumble that they want to use their self-driving car when they head to the hills, or the lake, or when they are moving. The answer that an AI self-driving car can't do towing won't be satisfying. Worse still, forcing or tricking an AI self-driving car into doing towing, if it doesn't realize what's taking place, I'd say is a formula for disaster. If we can get to the moon, seems like we can get an AI self-driving car to be relatively safe and sound as a towing driver. We're aiming for that to happen.

CHAPTER 4
DRIVING STYLES
AND
SELF-DRIVING CARS

Lance B. Eliot

CHAPTER 4

DRIVING STYLES
AND
SELF-DRIVING CARS

Have you ever been behind a slowpoke car?

It always seems to happen that there's a car going excruciatingly slow when you are in a rush to get to your destination. You try to look through the rear window of their car to get a sense of what kind of person is in the driver's seat, and you are curious about why that person can't seem to find the accelerator pedal. Your first thought is that maybe it's a senile person that no longer seems to be able to properly control a car. Or, maybe it's a young timid driver that worries they'll get a ticket for going to fast. Those fully compliant stops and long pauses at a stop sign are enough to make you want to honk your horn and maybe even shove the car ahead of you out of the way.

In driving parlance, those slow drivers are often referred to as "dead time" drivers. They aren't in any rush. They have all the time in the world. For them, they believe that driving slowly is the right way to drive. If they see a yellow light up ahead, by gosh they are going to make sure to stop before it turns red. Trying to run a yellow light would be out of the question. On the freeway, they love to go slowly in the slow lane. Even more beguiling is that they have no hesitation to drive slowly in the fast lane, though they would prefer to stay out of the fast lane because it is normally filled with brutes.

Speaking of brutes, have you ever had a crazed fast driving driver that came right up to your bumper, even though you were already going well over the speed limit anyway. These speed demons believe that any open space on the roadway is a bad thing and that their car should be able to go wherever it wants. Running a yellow light is of course the only way to properly drive. Indeed, if the light goes red, well, as long as you aren't overly conspicuous then it's okay to run through a red light that just turned from yellow.

These fast drivers are often referred to as "deadline" drivers. They drive as though they are headed to the hospital and if they don't make it there on time then their spleen will burst. Urgency is their middle name. Stomping on the accelerator pedal is the preferred mode of driving a car. Some will say that they speed because they can, while others might be more deferential and merely blame the speeding on their car (sir, it's got a V8 engine made to go fast, what was I to do, they lament to a motorcycle cop).

These are two dramatically different driving styles, namely the slow driver and the fast driver.

The slow driver is considered to be the turtle or tortoise, while the fast driver is considered to be the eagle or the hare. A slow driver would assert that they are rightfully cautious, and we should all be happy that they exist. The fast driver would assert that as a fast driver they make our roads more efficient by getting to where they are going quickly.

When you ask each type about the other type, you'll get some harsh words from each of them.

The slow driver says that the fast driver is risky, foolish, pigheaded, dangerous, egotistical, selfish, and otherwise a lousy driver that's out of control. The fast driver says that the slow driver is an idiot, a dolt, should not be allowed on the roads, should take the bus, and otherwise is a lousy driver that's out of control.

Take a look at Figure 1.

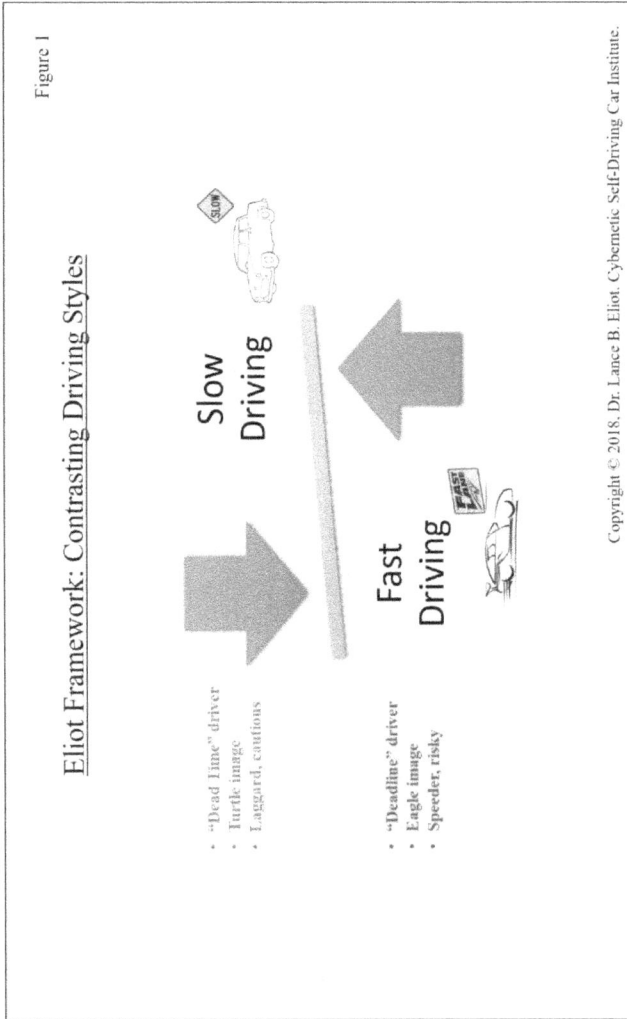

Figure 1

Eliot Framework: Contrasting Driving Styles

Slow Driving

Fast Driving

- "Dead Time" driver
- Turtle image
- Laggard, cautious

- "Deadline" driver
- Eagle image
- Speeder, risky

When you mix together the slow driver and the fast driver, it can be an explosive mixture. The fast driver, when bogged down by the slow driver, often will attempt even crazier driving maneuvers than usual, including the other day I saw a speeding driver that opted to swerve into the oncoming lanes to get around what they perceived to be a slow driver. It's a recipe for car crashes, for sure.

Take a look at Figure 2.

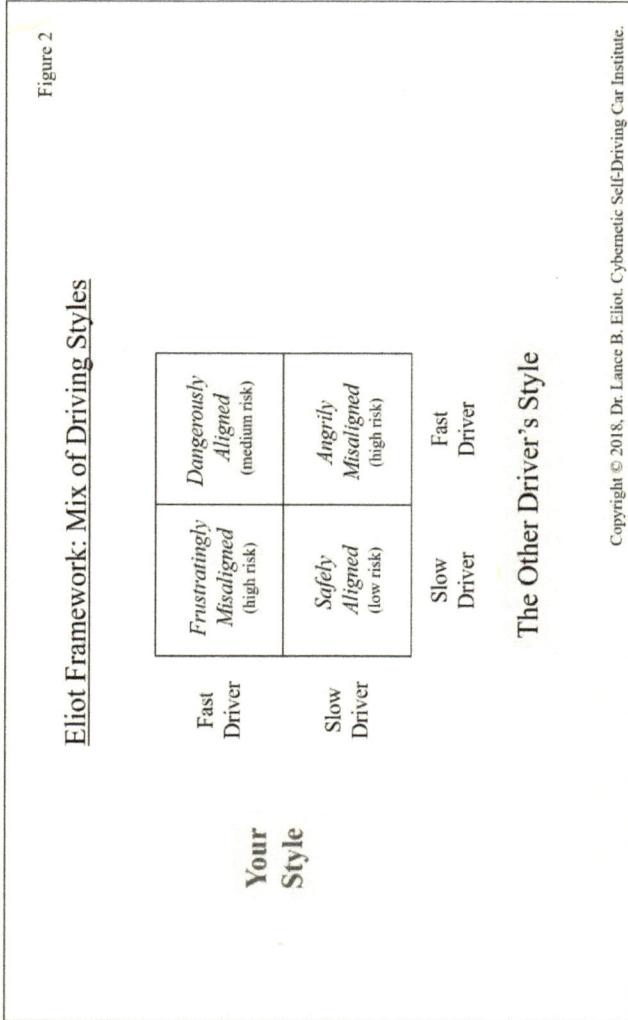

Figure 2

Eliot Framework: Mix of Driving Styles

	Slow Driver	Fast Driver
Fast Driver	*Frustratingly Misaligned* (high risk)	*Dangerously Aligned* (medium risk)
Slow Driver	*Safely Aligned* (low risk)	*Angrily Misaligned* (high risk)

Your Style

The Other Driver's Style

I've shown a four square that consists of the fast driver and slow driver perspectives. On the left side, I've listed the fast driver and the slow driver, so pick whichever one you are (or, if neither, then just take a look at each perspective). On the bottom of the four square I've listed also the slow driver and the fast driver, though this is with respect to "the other" driver (whichever other driver you might so encounter).

The fast driver perceives the slow driver as frustrating and this creates a high risk driving circumstance (see the upper left quadrant). The slow driver perceives another slow driver as equally aligned with them so it is considered a relatively safe encounter (the left lower quadrant). The slow driver that encounters a fast driver is bound to get angry at the fast driver, such as by getting cut-off by the fast driver, and so this circumstance also creates high risk (lower right quadrant). When the fast driver encounters another fast driver, it can be dangerous, but I'll call it as medium risk, though admittedly it could readily become high risk if they opt to compete with each other (upper right quadrant).

Besides the overall discomfort that is created by a slow driver or a fast driver, there's also the question of what's ill-advised and also what's just outright illegal in terms of driving.

Let's consider some examples.

In the first example, let's assume we have a slow driver ("S") that is driving so slowly that it is ill-advised and has a tendency to confound other traffic. There's a fast driver ("F") that is also driving in an ill-advised manner, likely allowing insufficient braking space between them and the car ahead of them. In the second example, the slow driver is driving fully in a legal manner, while the fast driver is driving in an illegal manner, which might consist of exceeding the speed limit and making maneuvers that endanger other cars and pedestrians. In the third example, there's a slow driver that is just on the edge of being illegally driving, and a fast driver that's driving in an ill-advised manner but has not yet verged into the illegal zone.

To some degree, the notion of "slow" and "fast" driving is somewhat relative. Two cars could be going the proper speed limit, and yet one might be considered a slow driver and the other a fast driver. Within the latitude of legal driving, you can go slowly, and you can go fast, doing so without necessarily violating any laws. I've known fast drivers that when they encountered another fast driver it was as though the first fast driver was a slow driver in comparison. I've seen a slow driver that encountered another slow driver, and then loudly complained that the other driver was exceedingly and frustratingly too slow.

Take a look at Figure 3.

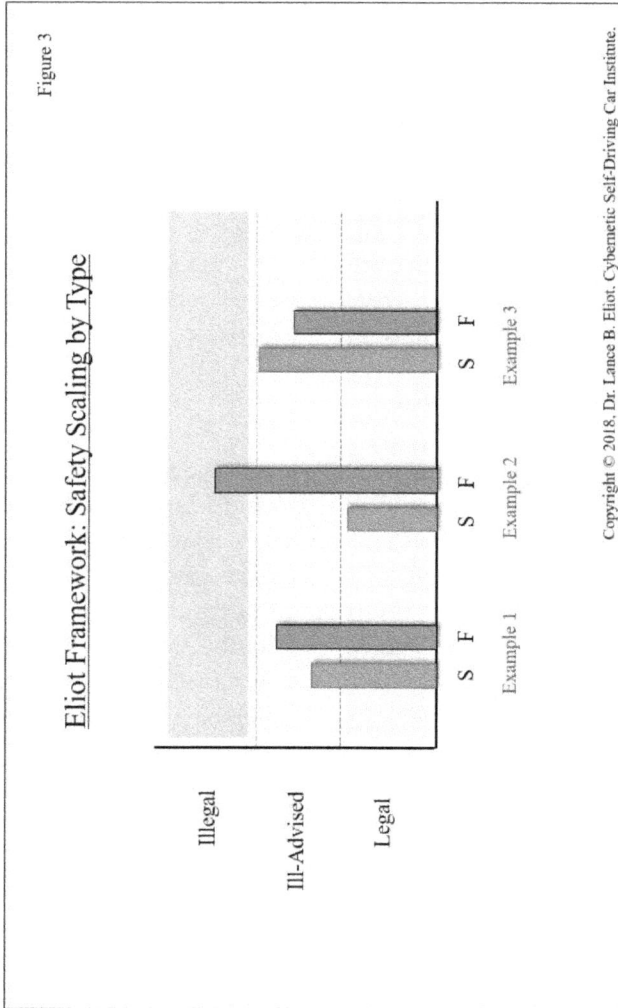

Figure 3

Eliot Framework: Safety Scaling by Type

Copyright © 2018. Dr. Lance B. Eliot. Cybernetic Self-Driving Car Institute.

That being said, most states have as a provision that any kind of driving that is considered inappropriate can be labeled as illegal if it endangers those on the roadways. You can be driving the speed limit, and even less than the speed limit, and still get a ticket, such as if it's a rainy day and the streets are slick, so much so that a police officer believes you were driving recklessly. There are some that drive fast because others around them are doing so, and likewise some that drive slowly because other cars around them are doing so. In that sense, peer pressure can impact our driving style.

Circumstances can impact whether you opt to be slow driver or a fast driver. You might normally be a "medium" driver that generally doesn't go too fast and nor too slow. But, on a given day or time, and given location, you might switch over into another style. Suppose you are stuck on the freeway and its pure bumper to bumper traffic. It's hard to be the fast driver in that situation and so you're more likely to momentarily be the slow driver. If you are on the open highway heading across the country, and there's no other traffic for miles, you are likely willing to become the fast driver.

Let's then agree that there are some drivers that are generally medium, there are some that are generally a fast driver, and there are some that are generally a slow driver. That's their dominant style. This style might vary depending upon a given circumstance. If the situation dictates, any of the three styles might switch into one of the other two styles. Usually, though, their dominant style will account for the bulk of their driving.

Now take a look at Figure 4.

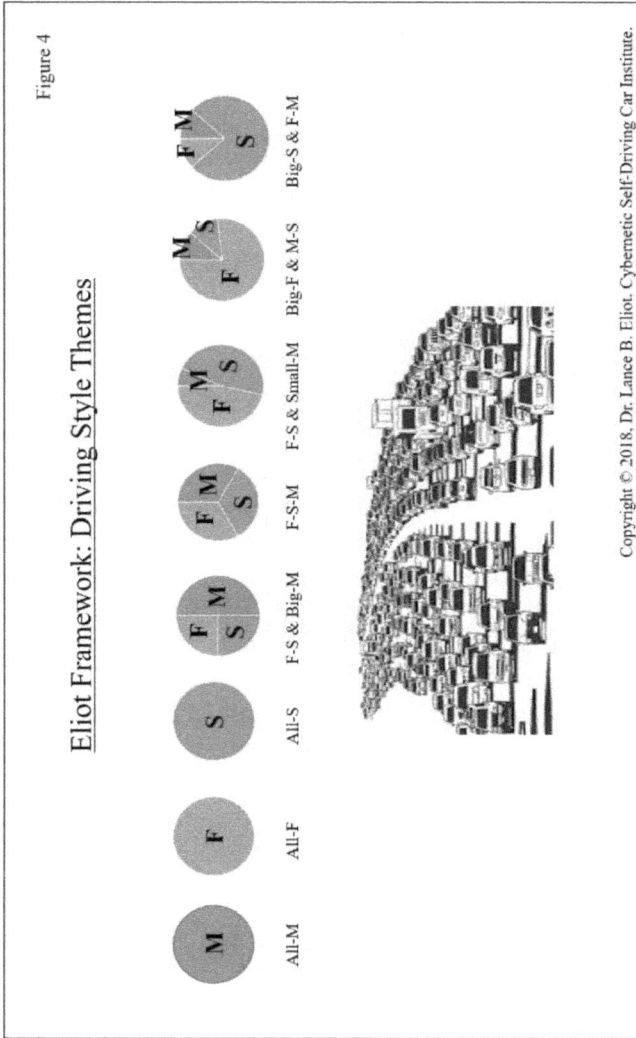

Figure 4

Eliot Framework: Driving Style Themes

All-M | All-F | All-S | F-S & Big-M | F-S-M | F-S & Small-M | Big-F & M-S | Big-S & F-M

Copyright © 2018. Dr. Lance B. Eliot. Cybernetic Self-Driving Car Institute.

For any given population of drivers, P, we could have that all of the population consists of entirely Medium drivers (P=M). Or, we could have a population of all fast drivers (P=F), or perhaps all slow drivers (P=S).

It would be somewhat unusual though to have a large population of drivers that was so homogenous. The odds are higher that the population of drivers will consist of a mixture of M, F, S. For example, as might have a population of mainly M (Big-M), and the rest are split between F and S. We might have a population that's an equal one-third of M, S, F. And so on.

What does this all have to do with AI self-driving cars?

At the Cybernetic Self-Driving Car Institute, we are developing AI software that enables a self-driving car to be M or S or F, and be able to shift their driving style as needed, and also be able to detect the style of other cars around them.

Right now, most of the existing AI self-driving cars are relatively slow drivers and you can properly suggest they should be classified as embodying a S style. The auto makers and tech firms would rather have their self-driving cars walk before they run, so to speak. The moment that your self-driving car starts to get up into higher speeds, it means there's less time for the AI to react and a greater chance of the self-driving car getting into untoward situations.

Allow me to clarify. If an AI self-driving car is on the freeway, it's going to go pretty much at the speed of the surrounding traffic and abide by the speed limit. In that way, it's not especially a slow driver. On the other hand, most of the freeway traffic that I encounter is almost always going faster than the speed limit. Almost none of the self-driving cars are programmed to do this. Thus, by going at the speed limit, and if they are on the freeways I'm on, I assure you that the rest of the traffic would consider that self-driving car to be a slow driver.

I've had some AI developers from some of the auto makers that tell me that those human drivers going faster than the speed limit are wrong and driving illegally. As such, these AI developers don't seem to care that the law abiding self-driving car is actually creating a dangerous traffic situation by going slower than the prevailing traffic. It's the stupid humans that are at fault. Though I certainly understand their viewpoint, it nonetheless does not recognize the reality of the driving world.

Human drivers that encounter a self-driving car are right now pretty much willing to give the self-driving car some wiggle room and not be bothered by the slow driving aspects. It's akin to driving near a teenage novice driver. We all allow those drivers a lot of room and figure that their aren't many of them, so why not just let it be. There are so few self-driving cars on the roadways that if you do encounter one, it's a novelty and you are glad to let it drive however it wants.

Imagine though if you had a thousand of the teenage novice drivers all surrounding you while you drove. Your leniency toward them would wear thin. I've predicted many times that once we get a lot of self-driving cars on the roadways, the "polite" human drivers that are right now allowing self-driving cars to do what they want to do, will become a lot less polite. Being boxed in while on the freeway, by perfectly law abiding "slow" self-driving cars to my left, right, ahead, and behind, well I'm betting that not all human drivers are going to tolerate it.

Generally, I'd wager that most large populations of human driver traffic consists of the theme of about half embodying a medium driving style, and a quarter are the fast and a quarter are the slow driving styles.

Meanwhile, right now, you could say that the AI self-driving cars are nearly all of a slow driving style, with some having occasions of fast and medium driving styles mixed in.

Suppose we have a circumstance of a bunch of human drivers that are of the fast driving style, and they come into contact with a bunch of self-driving cars that are of a slow driving style. This is likely to produce adverse results. A fast human driver is bound to ram into the back of a slow AI self-driving car, due to the fast human driver speeding along and getting caught off-guard by the AI self-driving car that's going slowly.

Take a look at Figure 5.

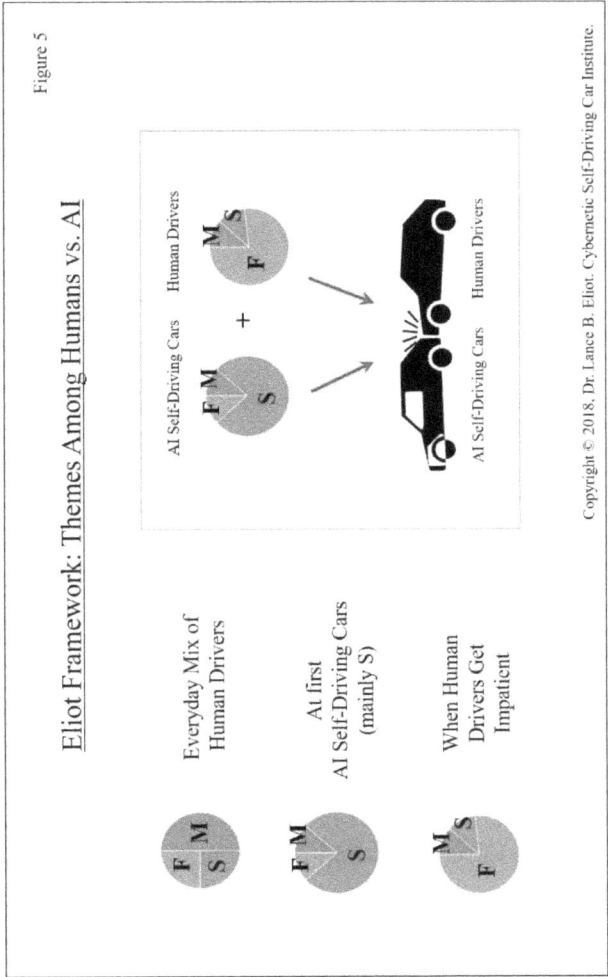

Figure 5

Eliot Framework: Themes Among Humans vs. AI

Copyright © 2018, Dr. Lance B. Eliot, Cybernetic Self-Driving Car Institute.

This is why we believe it's important for the AI of the self-driving car to be able to detect the driving styles of other drivers. Few of the self-driving cars today try to do this. Instead, the AI just acts in a myopic fashion of whatever car happens to be in front of it. If the car ahead goes fast, it doesn't have any significance to the AI, other than the now allowed space ahead of the AI self-driving car in case the self-driving car wants to speed-up.

Today's AI self-driving cars act as though the other cars around them are transitory transactions. We instead take the approach that the other cars around the AI self-driving car are creating a relationship with the self-driving car. The AI notices the driving nature of the other cars and builds up a track record about how those other cars are driving. The track record might be somewhat fleeting in the sense that perhaps the other cars around you are only there for a few minutes. The track record can be much longer such as if you are driving on a highway over a greater distance and you are jockeying with other cars that are nearby along the way.

By observing a track record, it is possible to then classify the other cars as to whether they are medium in driving style, slow in driving style, or fast in driving style. It makes a difference in terms of then predicting what those cars are likely to do. If you are following a slow driving human driven car, the AI can anticipate that it is likely that the human driver is going to take very cautious actions. Likewise, if the AI detects a fast driving style, it can anticipate that the human driver will take more risky actions.

In addition to detecting the driving styles of other cars, the AI is also programmed to be able to undertake a driving style as warranted for a given circumstance. It can be the medium driving style driver, it can be the slow driving style, and it can be the fast driving style. Each such style can be deployed when the situation makes sense to do so.

The model or framework for an AI self-driving car consists of the self-driving car using its sensors to detect other cars, it then does sensor fusion to bring together the sensory data into a cohesive whole, it then updates its virtual world model as to what the surrounding traffic situation consists of, and then it creates action plans as to what to do next, and then provides commands to the car controls accordingly. By labeling the surrounding cars that the sensors detect, and then tracking those cars throughout the virtual world model, the AI can characterize how those cars are driving and what their dominant driving style consists of, whether being M, S, or F.

During the formulation of action plans, the AI then can use that aspect to predict what the other cars around the self-driving car might do next. This continues over and over, and the AI is able to continually update the tracking to reflect what the other cars actually do. As mentioned earlier, a slow driver will not always necessarily drive slowly, and so the AI needs to be careful in not creating a myopic prediction.

Some AI developers have said that there's no need to do this kind of tracking because once there are all AI self-driving cars on the roadways, and no human driven cars, then there won't be the conventional M, S, or F anymore. Furthermore, the AI self-driving cars will communicate with each other via V2V (vehicle-to-vehicle communication). Once again, no need therefore to track and ascertain styles of the surrounding cars.

Yes, that utopian world is going to be something to see, but it's a long way from now. We have 200+ million conventional cars today in the United States alone, and those aren't going to overnight suddenly all become AI self-driving cars.

For the foreseeable future, we are going to have a mixture of human driven cars and AI self-driving cars. We need to develop AI systems to be able to cope with human drivers. For those AI developers that say it's up to the human drivers to cope with the AI self-driving cars, I'd say that's both narrow thinking and even worse it is thinking that's going to get a lot of people killed. We cannot assume that all human drivers are going to be willing to adjust their driving habits to whatever AI self-driving cars are put onto the roadways. Instead, we need to face reality and have the AI adjust to the human drivers. If this isn't done, you can bet that humans will revolt against AI self-driving cars and they'll (rightfully) insist to take them off the public roadways until the AI truly knows how to drive. That's a lesson we need to learn now.

CHAPTER 5

BICYCLISTS
AND
SELF-DRIVING CARS

Lance B. Eliot

CHAPTER 5

BICYCLISTS AND
SELF-DRIVING CARS

Are you living in Biketown or in Bikelash?

Let's start with Biketown.

Bicyclists, some would say, are wonderful because they are green, meaning they are good for society by using a non-polluting form of transportation. Many cities have opted to increase the number of bike lanes that they provide. Some cities even have specially painted traditional car lanes to indicate that those lanes are intended for bicyclists to ride in. A few cities have even removed car lanes entirely and opted to turn those lanes into bicycle lanes, plus sometimes also add a bit of greenery such as immovable planters.

Dockless bike-sharing services are now emerging as one of the hottest trends. The concept is that you can rent a bike, at any time, at any location, by simply seeing one within reach and being able to electronically unlock it, ride it wherever you want to go, and then park it wherever you want (the bike then electronically locks again and waits for another rider to rent it). No more having to keep a bike in a bike rack with a heavy steel lock on it. No more needing to own your bike. No more needing to go to a particular location where bikes are housed. Instead, bikes are like free ranging cattle. Via a mobile app, you can look to see where a bike is parked and then go there to start your ride. It's considered the "last mile" of ridesharing (you use a car-based rideshare to get near to a desired location, and then bike the remainder of the way rather than walking).

It's not all roses though in the biking world.

Let's consider Bikelash.

According to published statistics, there are an estimated 45,000 bicyclists injured each year in reported roadway accidents (that's the reported number, while the true full number including unreported incidents is likely much higher). The number of bicyclists deaths seems to range anywhere from 800 to 1,000 per year, and some numbers suggest that it really is more like 3,000-4,000 if you also include severe injuries that leave the bicyclist maimed for life. In short, anytime you get onto a bike, you've just increased your odds of injury or possibly death. Don't want to be sour on bike riding, and I'm just trying to emphasize that it's a dangerous "sport" and we often take it for granted.

As a quick note, the federal government prefers to call them pedalcyclists, which consists of riders of two wheeled non-motorized vehicles, tricycles, and unicycles that are all powered solely by pedals. I hope it's OK with you if I just refer to them overall as bicyclists.

Here are some fascinating numbers:

• 70% of bicyclist deaths occurred in urban areas versus rural areas (makes sense, density of traffic plays a role).

• 61% of the bicyclist deaths occurred at non-intersections (makes sense, usually drivers and bicyclists are a bit more alert while at intersections and watching for potential crashes).

• About half the fatalities were at night and about half during the day (you might find this at first glance surprising and might have assumed there should be more fatalities at nighttime, but it is probably reasonable to assume that there are many more bicyclists during daylight hours and less of a percentage that get killed, and probably though less numbers of nighttime bike riders they likely have a higher percent that gets killed).

- 96% of the bicyclists are killed in single-vehicle crashes (makes sense, all it takes is one car and one bicyclist to collide and the car is most likely going to survive while the rider does not).

- 84% of the fatalities involved the bicyclist getting hit by the front of the vehicle (makes sense, if a bicyclist rams into the back of a car they probably will be injured but not killed, while if the car rams into the bicyclist and likely doing so at a notable speed it's going to be bad times for the bike rider).

Who's at fault here?

Most of the bicyclists that I know would readily exclaim that it's the fault of the car driver. If the car driver had been paying attention, the car could not have struck the bicyclist. End of story. Their view is that no matter what the bicyclist was doing, there is no justification for the car hitting the bike rider. A car can always come to a halt, or swerve to avoid the bicyclist, or otherwise prevent the collision from occurring.

I don't want an army of bike riders to get mad at me, but I think this notion that it's all on the shoulders of the car driver is a bit over-the-top. I daily see bike riders that flout every known safety tip for bike riding. I say to myself, such-and-such is just asking to get hit. And even though, yes, a bike rider is legally considered a vehicle, I've said a million times that a bike is not the same as a car. Bike riders that think they are a car, are going to put themselves into dicey situations, and fault or no, the bike rider is going to lose this game of cat and mouse.

You might have forgotten that bicyclists are supposed to have the same rights and responsibilities as car drivers. We often times begin to think that bicyclists can just go where they may. In California, it's the law that bicyclists do these things:

- Obey all traffic signs
- Obey all traffic lights
- Ride in the same direction as traffic
- Signal when turning
- Signal when changing lanes
- Wear a helmet if under the age of 18
- Allow faster traffic to pass when safe
- Stay visible and not weave between parked cars
- Ride as near to the right curb as practical
- Do not ride on the sidewalk unless legal exceptions allowed
- Make left turns in the same way cars do
- Make right turns in the same way cars do
- At nighttime must have a front lamp
- Must have a rear red reflector or equivalent
- Reflectors on each pedal
- Etc.

When my children were first learning to ride a bike, I informed them about these above legal rules. Guess how long it took for them and their friends to abandon most of those rules? Not long.

Should we arrest every bike rider that does not obey the laws? Imagine how many arrests you'd need to make. The jails would be filled with bike riders. It would probably be the most prevalent crime committed. The number of police needed to catch and arrest all these scofflaws would mean we'd need to maybe double or triple the number of street cops. I suppose you'd have high school students with prison records going back to their days of kindergarten.

We can probably agree that we're not going to be arresting all of these unlawful bike riders. Can we get them to voluntarily be more lawful? There are attempts to achieve this goal, including some wonderful bike riding classes and local campaigns that tout being safe as a bike rider. Regrettably, these programs tend to change behavior only momentarily and then the bike riders revert back to their wild ways. It's hard to change behavior permanently in this sense, and it requires continual reminders.

What kinds of unlawful acts am I referring to, you might ask, well consider these:

- Tend to ignore traffic signs and blow through stop signs
- Treat traffic lights as a game that regardless of light color try to get through unscathed
- Ride in the opposite direction of traffic (quite popular!)
- Never signal when turning
- Never signal when changing lanes
- Be nearly invisible and weave between parked cars
- Ride sometimes near the right curb but really wherever judgement suggests
- Ride on the sidewalk (often done to avoid wayward cars)
- Etc.

I have to admit that riding in the opposite direction of traffic is very tempting. By doing so, you can see the cars coming at you. You have maybe a fighting chance of avoiding one hitting you. The problem with riding with the direction of traffic is that you can't see the car coming up behind you that is going to knock you off your bike and possibly kill you. I'm not going to argue here that we should change the laws about this, and I realize that driving facing traffic can be jarring for both the cars and the bicyclist. Just explaining why some people ride in the opposite direction of the cars.

A savvy bike rider is constantly watching how the cars are driving. Is that driver aware that a bike rider is nearby? Does the driver even car that a bike rider is nearby? Is that car weaving and maybe the driver is drunk? Is there a chance that one car will cut-off another car and the car so cut-off will weave into the bike lane? It seems like most car drivers consider "inconveniencing" a bike rider to be a small price to pay, and that it's better than possibly hitting another car or having another car hit them.

Unfortunately, not all bike riders are savvy bike riders. Also, some bike riders become complacent and after a while figure if they are still alive then they must be riding a bike correctly. Some bike riders don't know or don't remember what the rules of being on a bike are. There are also those bike riders that are determined intentionally to do unlawful acts and know they are doing so. It's their way of getting back at the man. This though seems shortsighted since if they get hit and killed, I'm not sure that they won over the man, so to speak.

I'd like to next shift focus to the car drivers in the equation of bike riders on-the-road and mixing with cars.

There are some car drivers that outright hate bike riders. I've seen some car drivers that purposely swerve their car towards a bicyclist. In other cases, they give the finger to bicyclist or roll down a window and yell at them. Get out of my way, they say. These drivers believe that bicycles should be banned, or at least forced to only be used in say parks or at the beach, in places where no car traffic is allowed anyway.

Recently, here in Southern California, when a local city decided to reduce the number of lanes in a particular stretch of road by making some of the lanes into bicycle lanes, the outrage became deafening once the change had been made. Drivers reported that they were now stuck in slow traffic. The nearby neighborhoods had cars roving through them, since the car drivers were desperate to get around the now constrained traffic. Shop owners said that less people drove to where their stores were located because the car drivers knew that the street was now chocked with traffic.

That generated a true bikelash.

Does this imply that all car drivers are angry at bike riders? No, certainly not. There are many drivers that are happy to share the world's roadways with bike riders. Unfortunately, what often happens is a few bike riders cause a problem, and the car drivers take this out on all bike riders. Likewise, the few car drivers that are especially mean to bike riders, cause many bike riders to become wary of all car drivers. It doesn't take much of a spark to cause car drivers to get fired up, and the same is true for bike riders. The rest of us are likely somewhere in-between. Nonetheless, us reasonable car drivers are surrounded by the car drivers that want to rid the planet of bike riders and we must contend with their antics.

Sadly, there are also the car drivers that seem to be living in their own bubble and rarely contemplate the plight of the bike rider. For these blind-deaf-dumb car drivers, they don't look for bike riders. They don't anticipate what a bike rider might do. They simply driver their car, straight ahead, and when a bike rider appears, it doesn't register in their minds, unless the bike rider happens to do something extraordinary. Often, at that point, it's too late for the car driver to do anything to avoid a collision. The daydreaming car driver can be just as dangerous to bike riders as the will-get-them-at-any-cost car drivers.

Speaking of costs, here's something else to consider. A car driver is typically wary of hitting another car. They are wary because they know that they themselves could get injured or killed. Subliminally, the average car driver does not think there's much of a consequence to hitting a bike rider. Yes, it would be bad. Yes, it might injure the bike rider. But, this is a lot less "serious" since the car driver is unlikely to themselves get injured or killed. The threat to bodily harm of a car-contacts-car is exponential in comparison to car-contacts-bike. Also, these drivers also figure that if they strike a bike, the bike rider is simply going to take a spill to the road, and maybe the bike gets a little bent up. No real damage involved. Car-to-car contact involves often significant repair bills to the car and a rise in car insurance.

That being said, any car driver that's ever been in an actual collision with a bike rider knows that this aforementioned concept is not what really tends to happen. The car driver can be injured or killed if in the nature of the collision they ram into something else in addition to the bike. For any driver with a conscience, the hitting of the bike rider will haunt them the rest of their lives. The injury to the bike rider can be severe and life limiting. The car driver can be charged with a crime. They can be sued to cover the damages. Thus, the "idealized" belief that hitting a bike is not that bad a thing, it's a whole different story once it happens.

What does this all have to do with AI self-driving cars?

At the Cybernetic Self-Driving Car Institute, we are developing AI systems for self-driving cars and included is the development of specialized software related to bicyclists, which by some of the auto makers and tech firms is considered an "edge" problem.

An edge problem is one that is outside the core of the overall problem being solved. Getting a self-driving car to properly drive down a road, being able to stay within the lanes of traffic, make turns legally, and otherwise drive like a regular car is supposed to drive – that's considered the core problem to be solved for AI self-driving cars. Having to deal with things like pedestrians, or things like bikes and bicyclists, well those are second fiddle and usually considered an edge problem. Definitely want to eventually solve an edge problem, but it's not the highest priority.

We believe that solving the self-driving car aspects of detecting and avoiding hitting bicyclists is a crucial aspect of being on the public roadways. A self-driving car that does not have provision for especially watching out for bike riders is about the same as the human driver that does not pay attention to bike riders. The head-in-the-sand approach will only last so long. Ultimately, inexorably, an AI self-driving car is going to hit a bike rider if there's no particular capability in the AI to avoid doing so.

I've seen some of the existing self-driving cars being tested on public roadways that don't have any bike riders present at all. We don't know for sure that those self-driving cars can handle dealing with bike riders. In other cases, there are bike riders present, but by stroke of luck the bike riders are dutifully abiding by the proper bike riding rules of the road. As such, once again the AI self-driving car can pretty much ignore them. The rule-of-thumb seems to be that don't bother me, I won't bother you. In other words, the AI self-driving car won't do anything to mess up the bike rider deliberately, and the AI self-driving car is hoping and betting that the bike rider will do likewise.

This does not take into account the bike riders that whirl and dance and go wherever they darned well please. The question arises as to what the AI will do with those bike riders. Some AI developers tell me that it's easily solved. If an object appears in front of the self-driving car, regardless of whether it is a bike rider or maybe a spaceship from Mars, all the AI has to do is detect the object and bring the car to a halt. It doesn't matter that it's a bike rider. The AI shouldn't need to care. Any object, the rule is, don't hit it.

Okay, I say, let's follow that logic along. A child is riding their bike. It's a school zone. The AI self-driving car is going the speed limit. We'll say it's going at 25 miles per hour (which is about 37 feet per second). The child, not paying attention to the car traffic, suddenly swerves in front of the self-driving car. The self-driving car needs to react. Can it come to a halt, having been going at 37 feet per second, in time to avoid the child that has nearly immediately appeared in front of the self-driving car? Answer, probably not.

Furthermore, maybe the self-driving car could have swerved to avoid hitting the bike. Or, maybe the AI should have been anticipating that a child on a bike might make an erratic action, and so have gone slower, maybe decreased speed to 5 miles per hour, as a precaution. Or changed lanes to give a wide berth for the bike rider.

A bike rider has certain characteristics that can be modeled and possibly predicted. The bike and bike rider are not just any object. They are not a light pole or a fire hydrant. They are a moving object. They have a particular kind of profile. We know that this moving object is intended to go in certain ways, and we also know that it can substantially decide to do something untoward.

Let's consider my framework for AI self-driving cars (discussed in Chapter 1). I'll walk you through the main elements of the framework as it pertains to this topic:

- Sensors

- Sensor Fusion

- Virtual World Model

- AI Action Plan

- Car Controls Command

The first aspect to consider is the sensor of the self-driving car. The hope is to be able to detect the presence of the bike rider. This can be potentially done via the visual sensors of the cameras. Imagine a picture of a street scene and you need to find the bike rider somewhere in the picture. This can be easy, if the bike rider is fully visible. This can be hard, if the bike rider is partially obscured by being behind another car or other objects. The visual aspects should be triangulated with the use of the radar, the sonar, and LIDAR (light and radar, if available on the self-driving car). Any of these sensors might catch a glimpse of a bike rider. The bike rider can appear and seemingly disappear, but hopefully at least one or more of the sensors is able to detect them.

Next is the sensor fusion. This involves bringing together the sensory data and trying to reconcile it. The bike rider might be detected by the LIDAR, but the camera can't spot him or her. Should the LIDAR be trusted or it is a false indication of a bike rider? The sensor fusion should be assessing which of the sensors is right or wrong, or at least potentially right or wrong. By combining together the bits and pieces from the multiple sensors, it possibly provides a strong indication of where the bike rider is.

During the virtual world model update, the AI should be tracking the bike rider. Where did the bike rider initially get detected? How fast is the bike rider moving? Is the bike rider riding smoothly or erratically? Does the bike rider seem to be a child or an adult? Does the bike rider pose a threat to the self-driving car? Does the self-driving car pose a threat to the bike rider? What can be done to reduce the risks of colliding with the bike rider? And so on.

From the updates of the virtual world model, the AI action plan needs to get updated. Maybe the self-driving car should slow down, and so the AI will be instructing the car to do so. Or, maybe alert the bike rider that the car is nearby and a danger is ensuing, this could involve honking the horn or taking some other conspicuous action. Or, speed-up. Or change lanes. Etc.

Finally, the AI then needs to issue commands to the controls of the car. This will then take time to be enacted. The AI will need to detect once the actual physical car has taken the action deemed needed, and then cycle back through each of these steps accordingly. In some cases, this will need to happen in splits seconds and so the timing of detecting the bike rider, predicting their actions, updating the model, updating the AI action plan, and issuing the car control commands can be crucial to avoiding a collision.

So far, the above highlights the acts of a solo bike rider. In real life, the odds are that wherever there is one bike rider, there will likely be more. It could be a school is nearby and a bunch of kids are riding their bikes to school. It could be a bike club and a gaggle of bike riders are out for their exercise. The point being that even though it seems like a hard problem to track and predict one bike rider, the odds are that this is a much more difficult problem because there are bound to be many bike riders all at once.

It becomes an interesting problem too to keep track of the various bike riders as though they are individuals. Allow me to explain. One approach is to just treat every bike rider as just another bike rider and happens to be here or there at a particular point in time. On the other hand, we human drivers often notice that say three bike riders are all riding smoothly, and there's a fourth one that seems to be veering outside the bike lane. Probably wise to keep an eye especially on the one that weaves outside the bike lane, we say to ourselves. Likewise, the AI should be virtually tagging the bike riders and trying to trace them over time. This is significant with regard to making predictions about their likely behavior.

There are moments at which the AI self-driving car needs to be acutely aware of the presence of bike riders. When getting ready to make a right turn, one of the more common mistakes is that a bike rider comes from the right of the car and the car turns directly into the path of the bike rider. I'm sure you've had this happen to you. You might complain that it was the "stupid" bike rider that caused this. Well, it would be better to try and have the AI self-driving car avoid hitting even a "stupid" bike rider, and so by being alert the AI can be anticipating it might happen and take steps to avoid a collision.

Another factor to consider is daylight and nighttime. Nighttime is going to be harder for the visual sensors of the self-driving car to detect a bike rider. Many bike riders do not have lights. This is a recipe for disaster. Inclement weather will also have an impact on the ability of the sensors to detect the bike rider. In short, the AI system cannot be programmed to simply assume that it will be nice and sunny, and that the profile of the bike rider will be one hundred percent noticeable.

There are also the use cases of a bike rider that is not actually riding their bike. Perhaps the bike rider is walking their bike. You might say this is then a pedestrian and no longer a bike rider. I'd suggest that it is more of a grey area. The walking person can suddenly hop onto the bike and start riding it. The AI self-driving car should be anticipating this possibility. The profile of a person riding a bike is also different looking than when riding a bike.

I mention the profile aspects because many of the AI self-driving cars use Machine Learning (ML) such as artificial neural networks for purposes of finding objects in visual images that are captured. The neural network is typically trained on thousands of pictures of people riding bikes. This then allows for the neural network to inspect a new image and try to gauge whether there is a bike rider in there. Suppose that the only pictures used to train the neural network consisted of riding bike riders. A walking bike rider then might not be detected as being a bike accompanied person.

Biketown versus bikelash. Bicyclists, love them or hate them. The AI self-driving car has to know about bicyclists since they exist and they are on the roadways. This edge problem is vital to becoming part of the capabilities of any proficient AI self-driving car. You could potentially have a Level 5 self-driving car that had no ability to detect and deal with bike riders (a Level 5 is considered the top of the scale and means that it is AI that can drive the car as a human can), but I would assert that such a lack in capability is not only a significant omission but I dare say not what we all would want a true self-driving car to be able to handle.

With a proficient AI self-driving car, there's a fighting chance to reduce the 45,000 annual biker injuries and the 1,000 or so annual deaths. Hold your breath for a moment when I say that if the AI isn't good enough, we might actually end-up with more injured bike riders and more human bike rider deaths. We cannot just assume that the AI self-driving car will magically eliminate those injuries and deaths. Bikelash will become AI-lash, if AI self-driving cars start hitting bike riders. Mark my words.

CHAPTER 6

BACK-UP CAMS AND SELF-DRIVING CARS

CHAPTER 6

BACK-UP CAMS
AND SELF-DRIVING CARS

One of the worst nightmares for any driver is the chance of backing up a car and running over someone. When you are backing up, it can be very difficult to know what's behind the vehicle. There's a famous video of a crawling baby in Brazil that unbeknownst to the parents crawled behind the family car as it was being backed out of the garage. The car got about halfway over the top of the baby when a person walking nearby pointed out there was a baby underneath. Getting out of the car, now stopped over the baby, the family members were fortunately able to pull out the baby and did so without much harm having come to the child.

They were lucky. Statistics were against them in the sense that by-and-large once a backout is underway, whomever is getting hit is likely to be severely injured or even killed. Per federal data, in the United States alone there are over 200 deaths annually and more than 15,000 persons injured via backover incidents. As you might guess, young children in the age of 5 or less account for nearly one-third of those deaths (of those, mainly children in the 1-2 years old bracket), while adults over the age of 70 are about one-quarter of the deaths. In essence, very young children and older elders are the most likely major segments of being the victim of a backover death.

I've been fortunate to never backover anyone, though I've had a few other close calls of varying kinds. In one case, while in college, a friend of mine thought it would be funny to hide behind my car as I was backing out, and then bang the car and act like I had hit him. I rushed out of the car and my heart was racing as I really thought I had somehow done this. My mind was instantly thinking about my first aid training and also where the nearest hospital was. When he stood up and laughed, I assure you that I did not consider it much of a joke and was darned angry about the whole thing.

I backed over some of my children's toys from time-to-time that they had left laying on the ground behind the car. I wised up to this and would always check behind the car before I started to back-up. They would get trickier though and sometimes the toys were already underneath the car before I started to back-up, and once I began going back I would hear and feel the crunch of some toy getting smashed by the tires or the underbelly of the car. In each case, it made me cringe because it reminded me that it could have been perhaps a human instead of a toy.

I'm sure that you are thinking that if we had backup cams on cars then we wouldn't have any of these deaths and injuries. Maybe.

First, you might find of keen interest that after years of delays in implementing a law passed by Congress in 2008 requiring regulators to enact legal measure that would require auto makers to enhance rearview visibility, now ten years later, in 2018, there is indeed a requirement that new cars being sold in the United States must be outfitted with backup cameras (the new requirement was announced in 2014 and the car makers were given four years to implement it).

For those of you with modern cars, you've already likely got a backup cam in your car, so this law doesn't mean much to you, other than the aspect that gradually there will be a lot of cars with the backup cam. Eventually, once older cars end-up on the junk heap, all cars will have backup cams as the newer cars become the dominant proportion of all 200+ million cars of today (this will take many years though to playout).

I guess we can call the backover problem solved since we're going to have all these backup cams – if you believe this you are in for a bit of a surprise. Turns out that a study in 2016 found that of cars outfitted with a backup cam and the same model of cars without a back-up cam, there was only about a 16% drop in reported backover incidents for those cars with the back-up cam. Think about that for a moment. You might have assumed that there should be a 100% drop in backover incidents. Having a back-up cam implies no more backovers.

Well, a backup cam is only as useful as the nature of the driver at the wheel. It is unlikely that all drivers will actually look at the display in their car to see what the backup cam shows them. For some people, they get so used to the backup cam that they rarely look at it. I know one driver that looks over his shoulder instead of looking at the backup cam display, which he insists is a better approach than relying upon the backup cam. Good old over-the-shoulder in his book is by far superior to the "useless" backup cam.

Even if the driver does look at the backup cam, they might not notice what the backup cam display is showing them. A baby laying on the floor behind the car might be laying still and thus there isn't any movement shown in the display, and thus the driver doesn't notice the child laying there. It might seem farfetched to you that someone would not notice a baby laying on the ground but imagine that you backout of your garage every morning to get to work, and some mornings you are in a rush, and 99% of the time there's nothing at all behind your car, and you "know" that the baby is inside the house (or so you assume). All of those factors can allow someone to mindlessly not see what the display is showing.

Another factor is whether an object moves into the field of vision at the last moment. Perhaps at first, before you start the car, you glance at the backup cam display, and mentally make a note that there's nothing behind you. So, you put the car in reverse and you take your eyes off the display. You look over your shoulder, or maybe in the rearview mirror, and begin to backup. You are feeling confident that there's nothing directly behind you. Using the story of the Brazilian baby that crawled behind a car, imagine if the baby was at the sides outside the view of the cam initially, and managed to crawl behind the car at just the most inopportune moment. This could happen in a few split seconds of time.

Of course, we also need to consider the field of vision of the backup cam as key to this too. Depending upon what kind of backup cam you have, and how it is mounted, you can have a narrow view or a wide view. You can have a view that sees down to the floor, or a view that is more of an upward look. There can be blind spots that the backup cam does not show you. The backup cam can also get obscured with dirt or other obstructions. Making assumptions that the backup cam gives you an all-knowing all-seeing vantage of what's behind the car is a mistaken belief.

Furthermore, some backup cams allow you to have a variety of vantage viewpoints, and you as the driver need to select the one that you want to see. Some drivers, being lazy or just not attune to the multiple views, have a tendency to leave the display set to a particular view. The driver doesn't rotate through them each time they do a backup operation. Most drivers become complacent about backing up when in a familiar setting. If you backup each day at your driveway or garage, you become accustomed to doing so. One sobering statistics is that by-and-large the person injured or killed is a family member or similar relation to the driver.

In essence, most of us will really only study the display of the backup cam when we get into dicey situations. You are at the grocery store and need to backup out of your parking spot. You saw that children were wandering around the parking lot and playfully having fun. All of a sudden, you devote your attention to the backup cam. Or, you are backing up and it's a really tight space situation, and so again you are on your alert and pay special attention to the backup cam display. The rest of the time, it's there but you don't put any mind toward it.

If the backup cam alone won't get the job done, I am sure you are thinking that let's put some automation onto the task. Indeed, there are some backup cam systems that have an alerting feature. If the backup cam detects an object in the field of view, it will make a tone or some other alert inside the car to let the driver know. This could definitely help for those drivers that aren't rigorously always studying their backup cam.

In addition to an alert, or in lieu of an alert, another kind of automation is an emergency braking system for backover prevention or mitigation purposes. If the emergency braking system detects an object within the field of view, and if the car is backing up and in motion, the emergency braking system acts like a collision avoidance system and stops the car, doing so regardless of what the driver might try to do. Automatic emergency braking systems while in forward motion are increasingly becoming common on new cars and will become an auto industry recommended requirement for all new cars and light trucks sold in the United States starting in 2022. This though does not apply to backup emergency braking systems.

Here's then the story so far. We know that backovers are a deadly problem. We know that having a backup cam helps to reduce the number of backovers, but doesn't curtail it entirely. We know that if you had an alert coupled with the backup cam, it would likely further reduce the backovers. If there was also an emergency braking system that applied to backing up, it too would likely even further reduce the backovers. As they say, we have the means, but we don't quite yet have the willpower. There is a cost to putting in an alert system and an emergency braking system for backover purposes, and society hasn't reached a point where it wants those so much that it has demanded that they be put onto cars.

What does this have to do with AI self-driving cars?

At the Cybernetic Self-Driving Car Institute, we point out that by-and-large self-driving cars are going to be equipped with sensors that allow for looking behind the car, and so it is an already built-in capability that just needs to be leveraged by the AI of the self-driving car. That's an "edge" problem that we are working on.

An edge problem is considered one that is not at the core of something. The core of an AI self-driving car is having the AI be able to drive the car forwards, and be able to drive down streets, drive on freeways, make safe left turns, and otherwise do all the things a human driver can do. The true self-driving car is considered a Level 5, meaning that it is a self-driving car that is driven entirely by the AI without any human intervention, and that the AI can drive as a human could.

In essence, right now, the auto makers and tech firms are mainly concerned with making an AI self-driving car that can drive forwards. Going in reverse is considered a secondary problem, or what some refer to as an edge problem, since its not at the core of the driving task (as they view it). Yes, it is an important part of driving, but it's not as crucial as driving forwards, they contend. In fact, some of the AI developers consider going in reverse to be fully solvable by just taking what you've developed for going forwards and reapplying it when the car is in reverse.

We don't believe that going in reverse is merely the same as going forwards.

This also brings up the important point that if you already are wanting AI self-driving cars so as to reduce the estimated 30,000 annual deaths in the United States due to driving incidents by human drivers, you would presumably also like to see that the 15,000 per year injuries due to backing up would also get reduce too. The AI self-driving car pretty much should already have the needed sensory devices, and so the other aspect is the AI software to leverage those sensors.

That being said, not all of the emerging AI self-driving cars necessarily have a typical backup cam per se. They might have other cameras on the rear of the vehicle, though not necessarily of the type for a backup purpose and nor aimed at the ground behind the vehicle. Some of the cameras are instead aimed at a further distance, so as to detect a car behind the self-driving car. One added potential plus for an AI self-driving car is that there are usually radar, sonar, and LIDAR on the car too, which can be used then in combination with the cameras.

I want to point out that very important element that I just mentioned. A conventional car that is outfitted with a backup cam is unlikely to have radar, sonar, LIDAR, and other sensors that can be used in combination with the backup cam. A conventional backup cam is all alone. It is the only means to try and detect what's behind the car. This is slim. Having only visual clues about what is behind a car can be misleading or distorted. It is advantageous to have multiple ways to detect what's behind the car.

We're incorporating the other sensors into the gambit of preventing backovers so that we can increase the chances of avoiding a backover, doing so by bringing together the visual data, the radar data, the sonar data, the LIDAR data, and the rest. The AI of the self-driving car has to be doing some solid defensive driving when backing up.

It is helpful to consider the major Use Cases associated with backing up a car. We back out a car in relatively common circumstances. There are exceptions beyond the common circumstances, but its best to focus initially on the common ones and then branch out from there.

First, there is the backing out of a garage as a driving-in-reverse type of task. This is extremely common for people to want to do. The task is not as easy as it might seem. Sometimes a garage has a lot of junk in it and the sensors cannot detect anything distinctive (it's one large blur). A garage can have very tight quarters and it makes the sensors unable to work appropriately. Garage parking and getting in and out is a specialized kind of problem (another edge problem!).

Next, there's the act of driving down a driveway while backing out. A similar use case is driving up a driveway while backing out. There's backing into a parking spot as another commonly performed task, and likewise backing out of a parking spot (less frequent, but sometimes combined with going back-and-forth to inch out of a tight parking spot).

Throughout any of those backing up operations, the AI needs to be on the watch for obstructions. The obstructions might be large or small. They might be still or in motion. They might be readily detected by multiple sensors, or only detected by one of the sensors. They might be moving away from the self-driving car or toward the self-driving car. It could be an obstruction that is recognizable, such as perhaps detecting that the obstruction is a person, or it might be an obstruction that is unrecognizable, which is nonetheless an obstruction but one of unknown capabilities or purposes.

There are also some important exceptions. For example, suppose there is a person purposely standing at the back of my AI self-driving car that is warning other people to stay clear. This helpful person, they themselves now become a detected obstruction by the AI. How will the AI know that the obstruction is actually part of the backing up operation? Instead, the AI is going to assume that the person standing there is to be avoided and will likely bring the car to a halt.

Speaking of which, every morning there is a newspaper tossed onto my driveway. Each morning, I dutifully back down the driveway and go over the newspaper. When I get home at night, I once again drive over the newspaper, and upon parking my car in the garage, I get out of my car and go to get the newspaper on the driveway. Let's for the moment assume that the self-driving car sensors and the AI are good enough to be able to detect that the newspaper is sitting there on the driveway. The AI would presumably refuse each morning to backup and later when I get home would refuse to go forward, since in both cases I am driving over something.

Thus, this is a harder problem than it might seem, since there are going to be circumstances where the human occupants want the AI self-driving car to proceed with backing up, in spite of a potential rollover of something or the nearness of a human or other object. The human occupant will need to have some means to communicate with the AI self-driving car, such as by using an in-car command. This human directive capability though has a downside, since suppose the human occupant intentionally wants to harm someone and so tells the AI self-driving car to proceed to backup into the person – should the AI self-driving car comply? As you can see, there are ethics issues involved in this too.

We know that backup cams are coming to conventional cars, slowly, gradually, and that some cars will also have alerts or emergency braking systems combined with a backup cam.

Older cars are unlikely to have this. Only some of the newer cars will have it. For AI self-driving cars, they are destined from the start to have sensory devices that can be leveraged for backing up safely. We just need to make sure that right kind of sensors are being included, and that the AI is savvy enough to leverage those sensors. I'd like to be able to say that with AI self-driving cars we'll have eliminated the backover problem, but realistically it won't eliminate it, but at least it should help to reduce the frequency and magnitude of backover incidents. Let's not back out of that goal.

CHAPTER 7

TRAFFIC MIX
UBER SELF-DRIVING CARS

CHAPTER 7
TRAFFIC MIX AND
SELF-DRIVING CARS

Car traffic can be downright exasperating, frustrating, beguiling, exhausting, and just a real pain in the keister (pardon my language).

Most of us dread getting stuck in traffic. Endless sea of cars. Stop and go movement. Bumper to bumper with a chance of some scrapes and fender benders. There are the looky-loos that aren't paying attention to the driving task and instead are looking at billboards or watching cows out in the fields (well, for more out-of-the-way driving locales). There are the drivers that seem to think that if they get within an inch of your bumper, it will somehow make the traffic go faster. There are the knuckle heads that have their blinker on, for miles at a time, apparently oblivious to the aspect that they are confusing other drivers and causing the rest of us to wonder if the dolt will suddenly change lanes without any actual viable notification.

Many us bemoan that it's not us that makes the traffic especially arduous, but instead those other drivers that seem out-of-touch or otherwise confused about how to properly drive in a traffic oriented situation. Normally, if the traffic is flowing smoothly, there are less of these disruptive drivers or at least they seem less apparent. Once the traffic starts to get clogged, it becomes a real survival-of-the-fittest world. Cars will swing into and out of lanes to try and get a few feet ahead of other cars. Drivers will raise a finger at another driver that appears to cut them off or try to jam into the lane ahead of them. In some cases, things can escalate out-of-hand and produce road rage with at time deadly consequences.

Generally, our existing laws and rules-of-the-road allow for human drivers to exercise a certain amount of discretion within loosely bounded legal rules. When a car to my right suddenly jams into my lane, and fails to signal, and fails to wait until there's a reasonable opening, and fails to clear my car by more than a fraction of an inch, you could say that they have violated the law by creating an unsafe driving situation. But, who's going to give them a ticket or stop them from this kind of discretionary driving? Unless by a stroke of luck there's a traffic cop there, this kind of driving behavior is going to be gotten away with, scot free.

Thus, the day-to-day traffic that we survive in can be characterized as not being uniform, and not necessarily enforceable as to the strict interpretation of the laws of driving, and overall allows for human judgement to be used to decide in what way someone will drive their car. Obviously, if the wayward drivers go beyond the allowed norm, and bash into another car, or they ram into a fire hydrant, or do anything of an "extraordinary" nature that's a demonstrative law breaker action, the odds are they'll get caught. I'm here trying to point out that at a level of driving just below that kind of clear cut legal outlaw stage, there is a lot of discretion as to how we can drive our cars.

If the roadway is pretty wide open and there are lots of lanes to choose among, the probability of a wayward discretionary driver that wanders up toward the legal outlaw line and getting caught is relatively low, while once there is a significant amount of traffic the probability generally gets higher. Also, once there is a significant amount of traffic, we now have more cars to contend with, and thus if say only 10% of drivers are the outlaw types, when you have just 10 cars nearby that means there is only 1 driver of the wayward type, while if you have 50 cars then you have 5 that are entered into the mix. In essence, volume of traffic makes a difference.

With the volume of traffic, we also need to consider density. Generally, the more volume and the higher the density of traffic, the more that the actions of one driver can impact the other cars and drivers. I realize this is not always true, in the sense that if you have a crazy driver even in light traffic situations, they can readily decide to weave in and around the other few cars and purposely try to cause them to hit their brakes or make other escaping maneuvers. It seems though that more often than not, the lighter traffic tends to seem allow for wayward drivers to do their thing without having as much a disruptive influence.

We could say that this is perhaps due to the magnitude of coupling. Cars that have lots of room around themselves to maneuver can be considered more loosely coupled to those other cars around them. When the space between cars is tightened, it tends to increase coupling. I am referring to a virtual kind of coupling, and not any kind of true physical connections. It's as though we had virtual invisible elastic bands that are connecting the nearby cars, and so when there is plenty of room then the cars are remotely coupled, and when they are closer to each other they are more directly coupled.

In close quarters traffic, if you suddenly get in front of my car by switching lanes without warning, and you do so within feet of my front bumper, I am likely to touch on my brakes to give you some added room and try to ensure that I don't ram into the back of your car. Meanwhile, the car behind me, which we'll say is also in close quarters, they too now might need to touch on their brakes. And so on it goes. A cascading effort can occur. It's like a bunch of dominos lined up.

If it wasn't close quarters traffic, and I touched my brakes, the car behind me which let's say in this circumstance is some distance behind me, might not need to touch on their brakes. Instead, via the natural flow of traffic, they might be able to come a little closer and then we are still moving along freely. When the density of the traffic increase, which usually is also associated with the volume of traffic, the coupling aspects tend to mean that whatever happens with one car can cascade into other cars. Again, think of dominos, but if they are spread out then when one falls it doesn't necessarily cause the other one to fall too.

There are times that I've been on the open highway and had a high density of traffic, even though the volume of traffic was low. A clump of cars happened to all get close to each other, even though the highway had miles upon miles of open space ahead of and behind us. This clumping can occur on a momentary basis, after which the cars then disperse out into the available open space. Or, sometimes the clump can remain a clump. This can be frustrating if you are part of the clump and don't want to be in it. You've perhaps had to slow down demonstrably to let the pack move ahead, or maybe you've sped up mightily to get way beyond the pack.

Some drivers don't seem to recognize when they've gotten themselves into such a clump. They are just keeping their noses down and staring at the car ahead of them. Several of these cars with the heads-down kind of driver can eventually can meet up. They become a clump, basically by default. Not one of them thinks to get out of the clump. Other cars come along, encounter the clump, and at times get stuck in the clump as well. It can be difficult to get around a clump and might take numerous adroit attempts to do so. The driving leapfrog techniques often applies when there are clumps.

What does this discussion about traffic mix have to do with AI self-driving cars?

At the Cybernetic Self-Driving Car Institute, we are developing AI systems for self-driving cars and have been studying extensively traffic mix and the nature of self-driving car driving techniques and approaches.

Some AI pundits claim that there's no need to study traffic mix since the world will be a wonderous place once we have all self-driving cars on the roadways. In this utopia, the self-driving cars will all communicate with each other via V2V (vehicle-to-vehicle communication), and politely share the roads with each other. An AI self-driving car will convey to another one that it wants to please go ahead and get into the lane with that one, and the responding AI self-driving car will communicate that yes, please do. They will dance together in a coordinated and cooperative manner. Furthermore, with the use of V2I (vehicle-to-infrastructure communication), these AI self-driving cars will be informed by the roadway that there's a bump in the road up ahead, and the AI self-driving cars will ready themselves to handle the bump. Nice!

This might someday be our future. But, until then, the real truth of the matter is that we're going to have a mixture of AI self-driving cars and human driven cars. I say this because right now in the United States alone we have about 200+ million conventional cars. Those conventional cars aren't going to disappear overnight. Instead, it will be years upon years, more like decades, before we gradually see AI self-driving cars becoming widespread and overtaking the number of human driven cars. We don't even know if eventually there won't be any human driven cars, and it could be that people will insist of still being able to drive a car.

AI pundits complain that if humans still insist on driving a car, it's going to mess things up. In one sense, they might be right. Based on studying the nature of traffic mixes, we can simulate what the future might be like in terms of the mix of human driven cars and AI self-driving cars.

Let's first start by thinking about proportions related to the mix:

- 1:N – this is one AI self-driving car that is in the midst of N human driven cars

- N:1 – this is N number of AI self-driving cars that are in the midst of one human driven car

- 10%:N – this is a ten percentage mix of AI self-driving cars in a volume of cars that includes N human driven cars

- 30%:70% -- this is a circumstance wherein the volume of cars has 30% that are AI self-driving cars and has 70% that are human driven cars

- Etc.

I'll be using this nomenclature when referring to the various mixtures in traffic of AI self-driving car and human driven cars. Let's also agree that when I refer to a volume of traffic, it is with respect to a given circumstance. I am also going to for now make the assumption that we have a relatively high density of traffic in these circumstances and the volume is relatively high too. I mention this for the same reasons that I had earlier stated that when the traffic is wide open, the nature of how the traffic intermixes is generally different than the circumstances when there is tighter coupling.

We also need to agree what we mean by an AI self-driving car. Herein, I'm going to refer to an AI self-driving car as one that is at the Level 5, which is a level at which the AI self-driving car is considered the driver of the car, and does not need a human driver, and can drive in whatever manner a human driver can drive. Levels less than 5 involve having a human driver available in the AI self-driving car, doing so in case the human driver needs to take over the driving task or is given the driving task from the AI.

It's important to consider that these traffic mix simulations involve a Level 5 self-driving car. For simulations with less than a Level 5, you would need to also include the aspects of the human driver that might take over the control of the self-driving car. This would definitely impact the nature of how the AI self-driving car is going to be reacting in the simulation since you'd need to account for the times when the human driver is driving the self-driving car versus the AI is driving the self-driving car.

Another factor to consider involves the sophistication of the AI self-driving car. Even once we get to the Level 5, there are going to be proficient AI self-driving cars at the Level 5 and others that we could reasonably agree aren't as proficient. Not all of the Level 5 rated AI self-driving cars will necessarily be at the same level of driving skills. Over time, there will be ongoing and continued improvements in the driving skills of even the Level 5 self-driving cars.

There are more factors to be considered too about the traffic mix situation. One really vital question involves how will human drivers react to AI self-driving cars? You might at first say that human drivers won't react any differently to an AI self-driving car than they do to another human driven car. Well, you'd be wrong. Human drivers will definitely be reacting differently to AI self-driving cars than they do to human driven cars, at least for the foreseeable future.

We've already seen that when AI self-driving cars are among conventional human traffic, the human drivers tend to give the AI self-driving car wide berth. It's as though the human drivers consider the AI self-driving car to be the equivalent of a novice teenager learning to drive. The human drivers are suspicious of the capabilities of the AI self-driving car. The human drivers are wary that the AI self-driving car might make an odd maneuver or do something untoward. Thus, human drivers opt to treat the AI self-driving car differently than a normal everyday human driven car.

You might object and say that the human drivers won't even necessarily be aware that another car around them so happens to be an AI self-driving car. I disagree.

Right now, the experimental AI self-driving cars on the roadway are at times rather obviously detected, since there are only certain brands of cars right now that are outfitted as self-driving cars. Also, noticeably absent is any driver in the driver's seat. Plus, some of the self-driving cars today have a LIDAR device on the roof of the car, and some also have branding on the sides of the self-driving car to point out that they are self-driving cars.

It is also relatively easy to detect today's self-driving cars by the manner in which they drive. Most of them are driving very slowly and cautiously. They come to a full stop at stop signs. They go less than the speed limit in places that most human drivers exceed the speed limit. They timidly proceed when an intersection light goes green. I realize that some human drivers also drive this way, but I am just saying that when you combine the driving behavior of today's AI self-driving cars with the other more physically apparent aspects, you can generally know when you are driving next to or near a self-driving car.

This is a crucial factor when creating a simulation. Most of the traffic mix simulations assume that the human drivers will be unaware that they are driving with AI self-driving cars around them. The simulations also assume that the AI self-driving car will drive in the same manner that humans drive cars, such as speeding, cutting corners, and so on. Or, worse still, the simulations assume that all drivers will all abide strictly by the rules of driving and be polite and respectful, regardless whether a human driver or an AI self-driving car. I think we can agree that human drivers don't drive that way.

Therefore, a realistic traffic mix simulation needs to consider for now that:

- Human drivers will drive as human drivers do, exploiting their allowed latitude and being wayward

- AI self-driving cars for the foreseeable future will drive in a more limited novice manner and be hardly wayward at all

- Human drivers will drive differently upon detecting that an AI self-driving car is nearby and will reactively drive because of the AI self-driving car being in their midst

Maybe, far in the future, human drivers will become accustomed to driving around AI self-driving cars and so they won't think twice about it. Likewise, perhaps in the future the AI self-driving cars will drive more akin to how humans drive, doing so with a bit of swagger.

Let's now return to the traffic mix proportions.

A recent research study seemed to suggest that having even one AI self-driving car, equipped with V2V, could improve safety and save energy in traffic (a study at the University of Michigan, "Experimental Validation of Connected Automated Vehicle Design Among Human-Driven Vehicles," partially funded by Mcity). I applaud the researchers for their efforts. Not only did they do simulated aspects, they also ran a series of experiments on public roadways with actual cars, including AI self-driving cars and human driven cars. This is the kind of work needed to help advance the AI self-driving car emergence.

In the experiment, the researchers were exploring what happens in a chain of cars when there is a cascading impact or chain-reaction due to a car braking and then re-accelerating. They found that the self-driving car was able to more smoothly deal with the circumstance, braking with 60% less of the G-forces and improving energy efficiency by 19%. The humans involved in the driving experiment were acting as a typical human driver might, namely tending to brake hard when caught by surprise about the chain reaction and then having to do a more rapid re-acceleration too.

Part of the trick here in this experiment is the V2V aspects. This is great, but it also is focused on the future of when we'll actually have widespread V2V. Until then, we're going to have AI self-driving cars on the roadways that either lack V2V, or are outfitted with V2V but no other cars anywhere near them also have V2V. Also, as stated in their research, they were focused on single lane types of driving, and more expansive studies are needed to look at a fuller mix of traffic including multi-lane situations.

In our simulations, using the aforementioned assumptions about driver behaviors, we've found that when you have the situation of 1:N, this tends to actually worsen the traffic situation. When there is a sole AI self-driving car among many human driven cars, it's the equivalent of having a novice teenage driver among many seasoned drivers. The novice driver tends to go slowly and react timidly, which then causes the seasoned drivers to become provoked and try to find ways around that driver. It's like a stream of water that a rock has been tossed into. The rest of the stream tries to find ways to get around that car. This is something to keep in mind in these early days of the adoption of AI self-driving cars on our public roadways.

In a similar kind of result, there's the N:1. When you have essentially all AI self-driving cars and mix into it just one human driven car, it tends to disrupt the traffic. This is because the wily human driver tends to drive in a wayward fashion, while the AI self-driving cars are all trying to work cooperatively and in coordination with each other.

Now, I would suggest that we're not going to see anytime soon an entire array of AI self-driving cars and one lone human driven car mixed together. By the time that happens, I'm betting that the AI self-driving cars will have been better equipped and programmed to handle the wayward human drivers and so they will be adept enough to cope with the human driver in their midst. In essence, I'm suggesting that by the time there is only a lone human driver among lots of AI self-driving cars, we will likely have first had a more proportionate mix of AI self-driving cars and human driven cars, and when the last few holdout human drivers are around will we have the N:1 circumstance (they'll have to pry the steering wheel from their cold hard hands, so to speak).

Let's consider the 30%:70% as an example. In this case, the simulation consists of a volume of cars in a high density situation that has 30% or about one-third that are AI self-driving cars, and the rest or 70% are human driven cars. What happens in this scenario? Under the same assumptions as earlier stated, the human drivers are now beginning to get used to the AI self-driving cars and adjusting their driving behavior accordingly. The traffic appears to sway toward the AI self-driving car mode of driving. It's as though the human driver are now among a sizable bunch of teenage novice drivers, and the seasoned drivers are giving into their way of driving.

We still see traffic waves occurring in this mix. Until the proportion of AI self-driving cars gets high enough and reaches a threshold, the human driven cars are still generally operating as humans do. With enough of the AI self-driving cars and once they get sophisticated enough, they are able to contend with those disruptive drivers. The disruptive drivers eventually too figure out that the AI self-driving cars are wise to them. Up until that point, the human drivers figure they'll pull the wool over the eyes of the AI self-driving cars and treat them like patsies that are readily exploitable.

Another aspect that we include in our simulations is the impact of other driving aspects such as the mix of having motorcyclists, pedestrians, bicyclists, and other traffic elements. Some studies focus on freeway only traffic situations, which then cuts out the pedestrians, bicyclists, and other city or street regular driving circumstances. Some

AI pundits say that we should have freeways devoted solely to AI self-driving cars, or if that's not feasible then at least have lanes dedicated to AI self-driving cars. The concept there is that we might be able to gain the advantages of the all-and-only AI self-driving cars by giving them their own place to drive. This though will require some potential hefty changes in our roadway infrastructure.

Today's traffic can be maddening. Imagine in the future when you get stuck in bumper to bumper traffic and look at the car next to you and there's no one driving the car. Will you still be able to show your finger to that non-human self-driving car? Even if you can, will it make a difference? In whatever manner this all plays out, I think we can assume that we'll have a small proportion of AI self-driving cars at first, which will gradually grow over time. At each of these stages of evolution of our traffic mix, we'll see somewhat different traffic patterns emerge. This is crucial to keep in mind when planning how we'll be dealing with the mixing together of human drivers and AI self-driving cars. Will it be like oil and water? Or, can we get it to be more like milk and cereal?

Lance B. Eliot

CHAPTER 8
HOT-CAR DEATHS
AND
SELF-DRIVING CARS

Lance B. Eliot

CHAPTER 8

HOT-CAR DEATHS AND SELF-DRIVING CARS

Have you ever seen a dog that was left in a parked car and the owner was nowhere to be seen? This happened to me a few weeks ago and I immediately got concerned about the welfare of the poor dog that was locked in a rapidly heating up car that was parked outdoors in the direct sun. There was about a half-inch opening in the passenger's side window which I am guessing the owner perhaps thought would be sufficient to provide fresh air for the dog. I pondered though what sane person these days does not already know about the potential dangers of leaving an animal in a car like this?

Other bystanders also gathered with me and we tried to figure out what to do. It was at a mall, and a mall security officer came over to find out why so many people had gathered together. When he saw what was happening, he explained that this routinely occurs and that at least a couple times a day there are circumstances just like this. He started to tell us what he would do if the owner did not show-up promptly, and by luck the owner did suddenly appear. The crowd was ready to lynch the owner. He got huffy at us and told us to mind our own business. I was crestfallen that he probably had not learned a lesson and likely would be repeating this same cruelty time and again.

I suppose we might give some latitude generally to people that do this and perhaps those people have a lack of awareness about the dangers of leaving a living being inside a locked car that is subject to the outdoor climate. In the case that I just mentioned, I no longer allow any latitude for this specific person because we told him straight out that he was doing something dangerous and should stop doing so. But, let's assume there are some people that really are the unwashed in the sense that they genuinely believe it is safe to leave a living being inside a car, unattended, locked in, and at the whim of the outdoor temperatures.

There is actually a greenhouse effect that happens inside a locked car that has the windows rolled-up. According to various charts, if the outside temp is 70 degrees Fahrenheit, a car can readily increase the internal temperature by about 20 degrees in 10 minutes. Thus, the internal temp is now 90 degrees. By 20 minutes, the temp will have risen another 30 degrees, and so it would likely be around 120 degrees inside that car. I realize that you can argue with these numbers by saying that the circumstances vary as to how high the temp would go and how quickly – all I'm trying to say is that the temperatures inside tend to get hotter than we think, and you can't judge this by the outdoors temperature. It could still seem to be a reasonable 70 degrees outdoors, and meanwhile inside that car it's a boiling 100 degrees or more.

Get ready for this scary statistic. The leading cause of non-crash car deaths for children under the age of 15 is heatstroke. Usually, a parent or a caregiver has left the child inside a locked car, which is parked outdoors, and the child is left in the car such that the greenhouse effect eventually kills the child. Horrible to even contemplate.

For those that think that leaving a small crack in the windows will prevent heatstroke, they are quite mistaken in this belief. The cracked open window is not effective to dissipate the heat. Get that out of your thinking. I've tried this myself in that I opted to sit in my car one time as an experiment, with the windows slightly down, and wanted to see whether it would make a difference. It did not. I was sweltering in due time.

A sad example of the dangers of this kind of situation involves a mother that had driven her 15-month old daughter to work, having placed the child into a backseat child-seat. On this particular day, the mother's normal routine had been thrown off and she left to work late and was rushed to get into the office.

Usually, the child was taken to a day-care but on that day the mother forgot about the day-care and drove straight to work in her haste. When arriving at the office, the mother parked in the parking lot and leapt out of the car to head into the workplace. The daughter was asleep in the backseat. The husband later in the day went to the day-care to pick-up the daughter, and when the daughter wasn't there he contacted his wife. The wife in a panic ran out to the car. I'll spare you the gory details.

It seems unimaginable to most of us that a mother would leave their child in the backseat of a car unknowingly. This is not just some piece of luggage or clothing that happened to be in the backseat. How could someone do this? Assuming it was completely unintentional, it would seem nearly impossible to have neglected to remember or notice that the child was still in that car. Analyses of these cases tend to reveal that the parent or caregiver was mentally distracted by some other matter, and often were outside of their normal pattern of activity.

In this case, apparently the mother thought that she had done the usual routine of dropping off her child at day-care. It happened each and every day. It was routine. She was outside her normal routine and got somewhat flustered, combined with a mental preoccupation about work. Her grief about this is immense and none of us can grasp the agony that she must live with every day since.

One technique that some say should be used involves always putting something "essential" into the backseat such as your wallet or purse. The odds of leaving behind your wallet or purse is relatively unlikely. You'll be forced somewhat to turn to the backseat and in so doing be reminded that your child is there. For the leaving of pets in a car, it is rare that someone forgets they left a pet in the car, and more the time is just sheer ignorance or outright foolishness to have left the pet in there. Certainly, there are some cases where a parent or caregiver leaves a child in the car to then be able to dash into a grocery store, and those too fall into the shocking aspects of ignorance or foolishness categories.

The act of doing these kinds of things are often subject to criminal laws. It is generally against the law to leave a living being in a locked and insufficient air circulating and temperature controlled environment. Society has wised up to this kind of behavior, and the "see something, say something" has helped to deal with the numbers of instances where this continues to recur.

Some have proposed that a device should be included into our cars to help prevent this from happening.

For example, a camera pointed inward that can "see" that there's a living being in the car, and then perhaps bleat a loud alarm or honk the horn, and possibly even be able to automatically unlock the doors. Maybe even have it be able to roll down the windows. Or, restart the car and turn-on the air conditioning. Instead of a camera, it could be a heat temperature gauge that is combined with a motion detector, and if there is motion inside the car and the temperature has gotten too high, and the car is parked, once again some kind of safety action is undertaken by the car.

These are all potential solutions to the problem. Unfortunately, almost no one is going to be willing to pay for these elements. It would raise the cost of a car. To retrofit a car with these elements will be somewhat costly too. Even if these all worked as prescribed, how many people would be willing to admit that they might someday leave a living being inside their locked car? I'd dare say that no one walks around thinking this is going to happen. I'll assume that most people assume that they will never ever have this happen to them. They are shocked to think that it could happen to them. They are not likely to beforehand go out of their way to purchase something to help prevent it from happening.

What does all this have to do with AI self-driving cars?

At the Cybernetic Self-Driving Car Institute, we are developing AI systems for self-driving cars and note something notable, namely that the technology that is already likely needed for a true self-driving car is going to be there anyway to help avert these situations, and so it's mainly up to the AI to be savvy enough to help solve this problem.

Most AI developers would say that this is an "edge" problem. An edge problem is considered at the edges of the core aspects of something that needs to be done, and as such those edges can be worried about at some later date. Focus on the core first, that's the motto. In the case of self-driving cars, the auto makers and tech firms are dealing with trying to get a self-driving car to be driven as a human would drive a car. The AI should be able to drive inside the lanes of traffic, it should be able to make right turns and left turns, it should be able to keep from hitting other cars and avoiding hitting pedestrians, etc.

Worrying about a living being left inside a parked car is just not very high on the priority list. Admittedly, with today's conventional cars, it doesn't happen so much that it is a true crisis of sorts. The volume of deaths and injuries due to heatstroke by being left inside a car is relatively small potatoes. When it happens though, it's sad and can have life-or-death consequences.

Let's also consider the future and see what might happen as AI self-driving cars begin to become available and pervasive. We are going to have parents that will put their child into an AI self-driving car and tell the self-driving car to take the child to day-care. The parent won't be riding with the child. Instead, the AI system is taking the child, alone, over to day-care. I know this seems incredible to contemplate, and I admit as a father it is unimaginable that I would do this, but I am sure this is the direction of society.

It's convenient for the parent. No need to drive out of your way. No need to leave work. Just have the AI self-driving car take care of your driving needs. This is predicated on the notion that the AI self-driving car is at a Level 5. A Level 5 self-driving car is considered an AI system that can drive the car as a human can drive a car and that there is no human intervention needed for the driving of the car. At the levels less than a Level 5, it is assumed that a human driver is available in case needed.

I am sure it will be years of proof that AI self-driving cars are at a Level 5 and of sufficient proficiency before people will trust putting their baby into the backseat and letting them ride alone. You might be willing to put a teenager into the car and figure that if somehow the self-driving car had an issue, at least the teenager could potentially take action. A baby or small child has not much chance of being able to take their own action. I'm not suggesting that these occupants could do any driving, only that once they are old enough, they could potentially get out of the car if they needed to do so.

I have often pointed out though that this ability to get out of the car is a dual-edged sword. You put your ten-year-old into an AI self-driving car. You tell the child to not get out of the car. Keep in mind that no matter how good the AI might be; the self-driving car is still a car. This means that the self-driving car is subject to mechanical breakdowns. Suppose the AI self-driving car breaks down on the freeway, half way to the desired destination with the child inside the car. Do you want your ten-year-old to sit inside that car, which presumably is now a sitting duck in the emergency lane of the freeway? Maybe yes, maybe no. Suppose that the child decides they are going to get out of the car on their own, under the thinking that it is safer, and steps out into the oncoming traffic? That's not good either. It's a conundrum.

As a society, we need to consider the societal and ethical implications of AI self-driving cars. Leaving these kinds of aspects up to chance, or letting the auto makers or tech firms try to decide, probably is not the best path.

There are some aspects that the technology of a self-driving car can be leveraged that might help in some means. I am not suggesting that technology is the total answer. Please don't infer that notion.

First, it is likely we're going to have cameras pointing inward in true AI self-driving cars. This will be pretty much built-in. If you are planning on using your self-driving car for ridesharing purposes, I'm sure you'll be happy to know that there's a camera that can be used to catch someone messing with the inside of your self-driving car. This also raises apparent privacy issues, but I've covered those in other columns and just note it as an important caveat herein.

The AI self-driving car is certainly going to be equipped to start the car on its own, drive the car on its own, and otherwise control the car. This is a given. As such, suppose that the camera detects that a child is in the backseat, and otherwise unattended, and if needed the AI could automatically start the car and start the air conditioning. This could either happen because the AI figured out to do so, or that the owner of the self-driving car was alerted and the owner perhaps then instructed the self-driving car to take such action.

We're also going to have some means of communicating with the AI of the self-driving car. More than likely, it will be a verbal interface akin to what we do today with Alexa and Siri.

The notion is that suppose an occupant wants to find out what the AI is doing, or wants to change the plans of the AI system as to driving the car, the human would talk with the AI. I might be inside my AI self-driving car, I tell it to drive me to work, and then along the way I see a Starbucks and so ask the AI to momentarily stop there for me to get some coffee.

If a child is in the AI self-driving car, it can potentially talk with the AI. The child might say they need help and ask the AI to call their parents. Or, maybe the child says they are bleeding and to go to a nearby hospital. Now, we'll obviously need to have some savvy AI because the child might say that they don't want to go to school today and to take them to the beach. It can't just be that whatever an occupant utters that the AI will blindly obey per se.

The AI of the self-driving car can also draw attention to the car, if needed, for the benefit of the occupants. Someone is in the self-driving car and they somehow get hurt, and so maybe the AI starts honking the horn or turning the headlights on and off to attract attention and aid.

Also, if the self-driving car has V2V (vehicle to vehicle communications), it can alert other nearby self-driving cars to come to the rescue by heading to wherever the self-driving car is parked (assuming you want this to happen). The AI could also call the police or the fire department, and even drive to a designated location to meet a police officer. Etc.

The microphone inside the self-driving car can be used to detect not only words, but other sounds such as a sound of someone getting hurt or in pain. Some self-driving cars will have an internal temperature gauge that comes with the normal operation of most cars anyway.

By using a combination of an inward pointing camera, temperature gauge, potential motion detector, audio microphones, and other such devices, the AI can possibly be a lifesaver for the occupants. I say can be, because if the AI is not developed to do this, it won't be happening by magic. Thus, as stated earlier, it's considered right now an edge problem and thankfully efforts like ours and others that see this as important are working toward solving it.

See something, say something, it's a motto for humans and for the AI of self-driving cars.

Once we begin having self-driving cars toting around our children, elderly, and other such occupants, I am betting we'll want the AI to be a helpful caregiver to watch over and protect those occupants. This has to be done in a manner that doesn't overstep what we as a society would want the AI to do.

Having an AI system that imprisons the occupants is bad, but at the same time "good" if it is done in a manner to protect the occupants properly and appropriately. Anyway you cut it, we want to avert the chances of anyone dying due to being left inside a parked self-driving car. Let's see if we can get the AI to help us on that sensible goal.

.

CHAPTER 9

MACHINE LEARING PERFORMANCE
AND
SELF-DRIVING CARS

CHAPTER 9

MACHINE LEARNING PERFORMANCE AND SELF-DRIVING CARS

My computer is faster than your computer.

How do I know this? Presumably, we would want to compare the specifications of your computer and the specifications of my computer, and by doing so we could try to ascertain which one is faster. This is not as easy as it seems and the result might not still convince you that my computer is faster than your computer.

Time for a bake-off!

We could have your computer and my computer try to go head-to-head on some task and see which one finishes the task sooner. The question then arises as to what task we would use for this purpose. Maybe I secretly know that my computer is very fast at making in-depth calculations, so I propose that we have the computers compete to see which one can produce the most number of digits of pi and set a time limit of say five minutes. Is this fair? You might know that your computer is optimized for doing text-based searches, and so you propose that instead of calculating pi, the two computers compete by doing text search in an encyclopedia to see which of the computers is fastest at finding a set of words that we come up with jointly.

If we cannot agree on what the task to use, we'll be hopelessly deadlocked as to how to determine which computer is the fastest. You are likely to keep pushing for tasks that fit to what your computer is fastest at, and I'll keep pushing for tasks that fit to what my computer is fastest at. We could somehow try to find a middle ground, but this might be hard to do and either one of us might think the other is somehow "cheating" with the task choice.

It would be handy if there was some kind of standard benchmark that we could use. Something that was chosen without specific regard for a particular computer per se, and instead something that we could use to gauge how fast our respective computers are. Plus, if it was a published standard, we could compare our results to the results of others that had also performed the task with their computers. There might already be performance scores about how fast various computers are related to the task, and so after we run the task on our respective computers, we could compare our run times with each other and with what the leaderboard says too.

For those of you that have been in the computer field for any length of time, you likely already know that this interest in computer benchmarks has been around for ages. Groups such as the Business Applications Performance Corporation (BAPCo), Embedded Microprocessor Benchmark Consortium (EEMBC), Standard Performance Evaluation Corporation (SPEC), and the Transaction Processing Performance Council (TPC) have promulgated all kinds of benchmarks. I remember fondly the Dhrystone, a benchmark developed for integer programming and which was named somewhat jokingly after another benchmark called the Whetstone.

Besides allowing comparison across different computers, these benchmarks can also be handy when dealing with the sometimes wild claims that are made by this vendor or that vendor. You've probably seen an announcement from time-to-time by some vendor that says their new processor can go faster than anyone else's processor. It's a claim that is easy to make. The proof would be had by having them run some standardized benchmark and then tell us how their processor performed. This can help others, whether academics pushing the boundaries of processors, or companies wanting to get the fastest new processors, since rather than simply believing the vendor's bold claims there would in addition be benchmark results to support (or refute) the claims.

What can also happen without having solid benchmarks is that it creates confusion and at times stymies progress in the computer field. If someone is thinking of buying a new computer and they hear that a new processor is coming out next month, they might be tempted to delay making their purchase. Suppose that the new processor is 10x the performance, it probably makes sense to wait. Suppose though that the new processor is about the same in performance, perhaps it makes sense to go ahead with the existing purchase. By having benchmarks, you could find out how the new processor fared on the benchmark, and have a clearer indication of whether to proceed or not.

Similarly, if a company or a researcher is developing a new processor, they would want to know what the marketplace already has. When working on pushing the boundaries of the hardware, maybe their research is going to just skyrocket performance and so they know that they'll have a home run hit on their hands. Or, maybe they realize by benchmarking that their new processor is just barely better than before, and so it might mean it's time to get back to the drawing board to rethink what they are working on.

Machine Learning Gets a Benchmark

My artificial neural network is faster than your artificial neural network.

How do I know this? Right now, it would be hard to prove it. We each might be making dramatically different assumptions about our neural networks. I could be running my neural network on a specialized hardware system that is optimized for neural networks and has 100 processors, meanwhile you might be running your neural network on your PC that has maybe four processors.

Besides running the model, there's also the data being used to train the model and also then exercise the neural network to see it perform. Suppose that I have used a dataset with 10 million training examples, and you've used a dataset that has only 10,000 training examples. Without some kind of agreement about the models we're both using, and some kind of agreement about the data that we're using, it's going to be dicey to make any kind of sensible comparison.

In the AI field of Machine Learning (ML), to-date there hasn't been an accepted standard of how to benchmark an ML system. This lack of a commonly accepted benchmark has had all the same downfalls as what I've earlier mentioned about the downfalls if we didn't have a benchmark for PC performance. Researchers can't readily compare their neural networks. Advances in neural networks cannot be readily compared. Companies that want to adopt neural networks are bombarded with claims from this vendor or that vendor about how fast their neural network is. And so on.

I am pleased to say that there is now a stake-in-the-ground for an ML benchmark. Called MLPerf, it is touted as "a common benchmark suite for training and inference for systems from workstations through large-scale servers. In addition to ML metrics like quality and accuracy, MLPerf evaluate metrics such as execution time, power, and cost to run the suite." (see the https://mlperf.org/).

Finally! We needed this. Everyone needs this. If you are an AI developer, you should be thankful that this now exists. If you are someone that relies upon an AI system using Machine Learning, this will be of benefit to you too. With a behind-the-scenes effort by high-tech ML-pursuing firms and universities, including Google, Intel, AMD, Baidu, Stanford, Harvard, and other entities, the MLPerf was officially released in May 2018. This is very exciting because from now on, whenever some new hardware is announced for Machine Learning, you can ask them how their fancy machine scored on the MLPerf. We seem to daily see announcements from Nvidia and Intel about their upcoming ML hardware, which henceforth we can ask for and expect to see scores related MLPerf.

For those of you that want to dig into the ML performance benchmark, it's all there in GitHub. Easy to find, easy to apply. No one can particularly complain that they couldn't find the benchmark, or that they couldn't get it to work. There have been occasions with other benchmarks where people tried to voice such complaints, and avoided using the benchmark because of it, but in this case it is hard to imagine anyone seriously being able to try that line.

When you consider the range of different ML kinds of needs and capabilities, formulating a benchmark is somewhat complicated since you want to cover as much ground as possible. So far, the MLPerf consists of 7 Machine Learning models and 6 ML-related datasets. They include ML for image classification, and ML for object detection, and ML for speech recognition, and so on. The MLPerf is certainly going to expand over time. Indeed, you are encouraged to consider providing additional benchmarks (more on this in a moment).

Here's the ML models and datasets, and they are categorized by Area and by Problem Type, there is also a Quality Target that anyone using the benchmark should be seeking:

Area: Vision
Problem Type: Image Classification
ML Model: Resnet-50
Dataset: ImageNet
Quality Target: 74.90% classification

Area: Vision
Problem Type: Object Detection
ML Model: Mask R-CNN
Dataset: COCO
Quality Target: 0.377 Box min AP and 0.339 Mask min AP

Area: Language
Problem Type: Translation
ML Model: Transformer
Dataset: WMT English-German
Quality Target: 25.00 BLEU

Area: Language
Problem Type: Speech Recognition
ML Model: DeepSpeech2
Dataset: Librispeech
Quality Target: 23.00 WER

Area: Commerce
Problem Type: Recommendation
ML Model: Neural Collaborative Filtering
Dataset: MovieLens-20M
Quality Target: 0.9652 HR@10

Area: Commerce
Problem Type: Sentiment Analysis
ML Model: Seq-CNN
Dataset: IMDB
Quality Target: 90.60% accuracy

Area: General
Problem Type: Reinforcement Learning
ML Model: MiniGo
Dataset: n/a
Quality Target: 40.00% pro move prediction

For those of you that are relatively new to Machine Learning, here's a handy tip – you ought to know each of the above ML models, and you ought to know the datasets that are being used with those models. When I interview to hire a ML specialist, these are the kinds of ML models and datasets that I expect them to walk in the door knowing (plus, other ML models and datasets as befits their particular specialty or expertise).

For those of you not familiar with ML, I'd like to point out that those datasets don't have to be paired to those ML models, it's just that those datasets are often used for those ML models. I say this because I am urging you to learn about each of the models, and also, separately, learn about each of the datasets. Keep in mind that they are distinct of each other.

If I was teaching a class on ML this coming year, I'd have the students become familiar with each of these, and I'd want them to be able to say what each does, its limitations, etc.:

- ML Model: Resnet-50

- ML Model: Mask R-CNN

- ML Model: Transformer

- ML Model: DeepSpeech2

- ML Model: Neural Collaborative Filtering

- ML Model: Seq-CNN

- ML Model: MiniGo

I'd also want them to be familiar with these commonly used datasets, in terms of what the data contains, how it is structured, how it is used for ML models, and so on:

- Dataset: ImageNet

- Dataset: COCO

- Dataset: WMT English-German

- Dataset: Librispeech

- Dataset: MovieLens-20M

- Dataset: IMDB

GitHub contains the code for each of the MLPerf chosen ML models, along with scripts to download the datasets, info about the quality metrics, and other characteristics to make things as transparent as possible. The ML models and datasets have so far been tested on a configuration consisting of 16 CPUs, one Nvidia P100, Ubuntu 16.04, 600GB of disk, and other related elements.

Take a look at Figure 1 as a handy overview.

MLPerf: Machine Learning Performance Benchmark

Figure 1

Vision	Area: Vision Problem Type: Image Classification ML Model: Resnet-50 Dataset: ImageNet Quality Target: 74.90% classification	Area: Vision Problem Type: Object Detection ML Model: Mask R-CNN Dataset: COCO Quality Target: 0.377 Box min AP and 0.339 Mask min AP
Language	Area: Language Problem Type: Translation ML Model: Transformer Dataset: WMT English-German Quality Target: 25.00 BLEU	Area: Language Problem Type: Speech Recognition ML Model: DeepSpeech2 Dataset: Librispeech Quality Target: 23.00 WER
Commerce	Area: Commerce Problem Type: Recommendation ML Model: Neural Collaborative Filtering Dataset: MovieLens-20M Quality Target: 0.9652 HR@10	Area: Commerce Problem Type: Sentiment Analysis ML Model: Seq-CNN Dataset: IMDB Quality Target: 90.60% accuracy
General	Area: General Problem Type: Reinforcement Learning ML Model: MiniGo Dataset: n/a Quality Target: 40.00% pro move prediction	*AI Self-Driving Cars (future expansion)*

Based on https://mlperf.org/ as of May 2018   Copyright © 2018, Dr. Lance B. Eliot. Cybernetic Self-Driving Car Institute.

If you are interested in the MLPerf, I recommend that you first read the MLPerf User Guide and then jump into the rest of the artifacts.

In the MLPerf User Guide, they point out that "a benchmark result is the median of five run results normalized to the reference result for that benchmark. Normalization is of the form (reference result / benchmark result) such that better benchmark result produces a higher number." And that "a reference result is the median of five run results provided by the MLPerf organization for each reference implementation."

This is important in that it is saying you need to do at least five runs for being able to indicate what your benchmark result is. Why do multiple runs? It could be that with one run you got lucky or there was a fluke, and so it is helpful to have multiple runs. When you have multiple runs, the question arises as to whether you keep all the runs and average them, or maybe toss out the high and the low, or take some other approach to trying to summarize the multiple runs. In this case, the MLPerf standard indicates to use the median.

Another important part of neural networks and ML models is the use of things like random numbers, which more formally refers to non-determinism. My run of a neural network and your neural network could each separately be materially impacted by initial seeded values or by other aspects relying upon random numbers. The MLPerf says that "the only forms of acceptable non-determinism are: Floating point operation order, Random initialization of the weights and/or biases, Random traversal of the inputs, and Reinforcement learning exploration decisions." They also indicate that "all random numbers must be drawn from the framework's stock random number generator." This attempts to level the playing field.

For training purposes, they do allow for the "hyperparameters (e.g., batch size, learning rate) may be selected to best utilize the framework and system being tested." This allows some reasonable flexibility in what the benchmark runners are doing, and could upset the playing field, but hopefully not (as I urge next).

I mention the above salient aspects because I want to bring up something now that we hopefully will all curtail, namely, there might be some that try to "trick the benchmark."

If you are determined to get a really high score on your new hardware or software for Machine Learning, you might be tempted to try and find loopholes in the MLPerf. You might figure that perhaps there's a means to get a higher score, even if your system is not really as good as the normal approach would depict. Therefore, you go through the MLPerf standard with a fine tooth comb, scrutinizing every word, every rule, every nuance, and try to find some miniscule rule or exception that you can exploit to your advantage.

That's not the spirit of things.

The MLPerf tries to prevent this kind of trickery from happening by saying this: "Benchmarking should be conducted to measure the framework and system performance as fairly as possible. Ethics and reputation matter."

It's difficult to try and write a standard that will cover all avenues of sneakiness. I am sure there are inadvertent loopholes in the MLPerf. The guide would have to be likely hundreds of pages long to try and layout every possible conceivable rule. If you find loopholes, it is hoped you'll let others know and that in the end we can plug them up or at least know they are there.

Machine Learning for AI Self-Driving Cars

What does all of this have to do with AI self-driving cars?

At the Cybernetic Self-Driving Car Institute, we are developing AI systems for self-driving cars that use Machine Learning and also aid firms in the assessing their ML systems and improving their ML systems. Having a Machine Learning benchmark is going to be handy.

The existing MLPerf does not directly address AI self-driving cars per se, though it does touch upon it. Yes, there is image classification, which is an important part of the sensor data analysis on an AI self-driving car. Object detection is another important aspect of an AI self-driving car such as finding a pedestrian in an image of a street scene. And so on.

What we need to do is have all of us in the AI self-driving car field provide added ML Models or added datasets into the MLPerf standard that are directly related to AI self-driving cars. The MLPerf welcomes submissions for expansion. They recognize that this initial release is nascent. The more, the merrier, assuming that whatever is added has value and provides a bona fide contribution towards the goal of having a solid base of ML models and datasets for performance benchmarking.

In the framework, it will be helpful to have proposed submissions into MLPerf for these major aspects of AI self-driving cars:

- Sensor Data Analysis for Self-Driving Cars

- Sensor Fusion Analysis for Self-Driving Cars

- Virtual World Model Analysis for Self-Driving Cars

- AI Action Planning for Self-Driving Cars

- Controls Activation Commands for Self-Driving Cars

- Strategic AI for Self-Driving Cars

- Tactical AI for Self-Driving Cars

- Self-Aware AI for Self-Driving Cars

In a movie starring Steve Martin, there's a famous set of lines spoken by him as the character Navin and another character named Harry, in which Navin exclaims: "The new phone book is here! The new phone book is here!" And Harry says, "Well I wish I could get so excited about nothing." Navin then replies: "Nothing? Are you kidding?!"

I mention this because the release of the MLPerf is something to get quite excited about. We can rejoice that an ML benchmark of sufficient quality and attention has been brought forth to the world. You can argue that maybe it's not complete, or maybe it needs tuning, or has other rough edges. Sure, that's all assumed. Let's at least move forward with it. Furthermore, I call upon my fellow AI self-driving car industry colleagues to find a means to add to the MLPerf with Machine Learning models and datasets that are specific to self-driving cars. I am betting it would be a welcomed contribution.

CHAPTER 10
SENSORY ILLUSIONS
AND
SELF-DRIVING CARS

CHAPTER 10

SENSORY ILLUSIONS
AND
SELF-DRIVING CARS

Are you a Yanny or a Laurel?

Unless you've been living in a cave, you likely know about the recent craze over an audio clip that has created a social media frenzy and sparked a debate among both friends and foes alike. A short audio clip that was posted on Twitter had asked listeners to report whether they heard the word "Yanny" or the word "Laurel" when hearing the clip. Thousands upon thousands of replies seemed to suggest that the world is split into Yanny believers versus Laurel believers. At times, it has been an even split, while at other times the tide starts to go toward the Yanny side and the next moment it slides over to the Laurel side.

I've overheard people talking about this curious and mind-puzzling audio phenomenon while I've been in the line at Starbucks, while at the coffee machine in the office, while at the grocery store getting my weekly foodstuffs, and even while camping in the middle of the woods. If a tree were to fall, would it sound like Yanny or sound like Laurel? There are now zillions of memes online about the matter and numerous clever take-offs have been crafted in both audio and video formats.

This seems reminiscent of the craze in 2015 that had everyone debating whether a picture of a striped dress was composed of the colors white and gold versus the colors of blue and black. It became an instant hit, it provoked lighthearted controversy, it lasted for a while as an intense focus of discussion worldwide, and eventually petered out.

Is all of this some kind of mass hysteria? Maybe the power of social media is such that it can cause people to go mad. Or, maybe people are so desperate for something novel or interesting that they latch onto these fads. The fascinating aspect is that people become quite agitated that whatever they think is the right choice, they are baffled that anyone could be on the other side of the issue. There was a Yanny proponent the other day that insisted that the Laurel proponent they were arguing with was purposely trying to provoke a controversy and merely pretending to hear the word as Laurel. This Yanny-fanatic was sure that there was a giant conspiracy going on, and that those on the Laurel side did not genuinely hear Laurel and were claiming they heard it for purposes of irking the rest of the world.

These kinds of debates also give scientists a moment in the spotlight. Some talking-head scientists right away suggested it was indeed all-in-their-mind as a psychological matter that showcased what people want to hear. In other words, if you wanted to hear Laurel, and you were presented with the audio clip, you'd think you heard Laurel, even though it was maybe actually being pronounced as Yanny. Likewise, if you wanted to hear Yanny, and then you heard the clip, you'd be convinced it said Yanny. This notion that people were being led down a primrose path did not seem to widely bear out though.

Yes, it's true that people can often be seeded to think a certain way. Yes, people are known for becoming cognitively anchored to something and it is often hard to get them to shift from their original anchor point. Those aspects though don't seem to account for the rather large number of people involved in this social experiment of the Yanny's versus the Laurel's. Instead, the scientific explanation that would seem to be the most plausible overall, and account for the largest segment of those immersed in this controversy, would be the aspect of sensory illusion.

Here, in the sensory illusion explanation, we consider the nature of the sound clip and note that it is very short in length and of relatively poor audio quality. In that manner, it is readily open to potential interpretation. Were the audio clip longer, such as an entire sentence, you'd perhaps have a better chance of ascertaining what it says, and likewise if the quality was higher and more distinctive it might be less likely open to multiple interpretations. It just so happened to be short enough and ambiguously sounding enough that it allows the ear to hear something that is not fully defined, and then the brain enters into the matter and tries to help clean up what was heard. This is akin to taking a blurry visual image and cleaning it up by refocusing the image and adding more pixels to it. Your brain is taking an ambiguous audio element and trying to make sense of it, doing so by internally polishing it and then trying to match what it heard to other sounds that it knows.

Have you ever been camping, and you looked off in the distance and saw a shape that maybe was a bear? Or, is it a human? Or, is it just a log that happens to be in the overall shape of a bear or a human. Or, maybe its bigfoot. But, anyway, the point is that your eyes can be tricked by visual illusions in that you see something vaguely and then your brain tries to polish it and match it to things that you know. Being at a distance of the image, you only have scraps of visual cues to work with. Your brain takes whatever morsels are available and tries to make it into something usable.

Airline pilots are known to be susceptible to visual illusions. There are many famous cases of airplane pilots that looked at a landing strip and thought that it was wider than it actually was, or shorter in length that it actually was, or that it was more upsloping than it was, or more down sloping than it really was. There are all kinds of visual illusions that pilots are supposed to be on the watch for. One is called the black-hole and it occurs typically when there is a body of water prior to where a landing strip is. If you've ever looked out a plane window when landing at an airport near the water, and at night time, you've probably looked down and observed that the water looks entirely blacked out. Rather than visually perceiving it as a body of water, it nearly looks like a mysterious black hole, as though the earth didn't exist there, and it was just wide open empty space.

Vection is an Illusion of Self-Motion

Sensory illusion can also include tricks of motion sensations. Sometimes, while sitting in bumper to bumper traffic, I'll notice the lane of cars to my right proceed forward slowly, and the lane of cars to my left proceed forward slowly, while my lane is at a standstill. This combination of motions to either side can occasionally create an odd feeling or untoward sensation that's called vection.

Vection is an illusion of self-motion. You believe that you are in motion, even though you are not. In the case of the cars around me, I at times perceive them as stopped, and I feel like my car is rolling backwards. It's a weird thing when it happens. If you've never experienced it, when you do so, you'll momentarily think the world's gone crazy and be vexed as to how in the world could your car be rolling backwards just out-of-the-blue. You might even reflexively stomp on the brakes of your car, doing so because your brain has told you to stop rolling backwards and the way to do so would be to bring the car to a halt. It's one of those bizarre illusions, for sure.[

If you are interested in car motion illusions, take a look at my article on that topic.

I am guessing that you likely accept the notion that there are plenty of optical or visual illusions that we humans can experience. And, that you likely also accept the possibility of motion-based illusions that we humans can experience. The idea of audio illusions is a bit harder for most of us to accept. You are tempted to believe that whatever is heard, it is heard in the same way, by all. But, if you think about foreign languages, and when you hear a foreign language that you don't quite know, I think you would agree that there are times when the foreign words are spoken that you might be unsure of exactly what you heard said. Your brain is trying to make sense of the sounds and at times it isn't sure what the sound really was. This can apply, perhaps incredibly, even to sounds that we think we know, such as the Yanny and the Laurel debate.

I'd like to take this Yanny versus Laurel debate and use it for another purpose herein, namely to spark discussion about the dangers of sensory illusions for a subject of another kind, as I'll explain in a moment.

What does this have to do with AI self-driving cars?

At the Cybernetic Self-Driving Car Institute, we are developing AI systems for self-driving cars, and are well aware of the dangers of sensory illusions that can impact the AI of a self-driving car. Auto makers and tech firms making such AI systems need to also be aware of the matter, and so does the general public that will be occupants in self-driving cars or otherwise be near to or around AI self-driving cars.

You might at first be bewildered by the possibility of sensory illusions being applicable to AI self-driving cars. The AI of a self-driving car is supposed to be automation that does not have the frailties of humans. Humans are the ones that are prone to sensory illusions, including the Yanny versus Laurel debate, and the examples I've mentioned about human pilots being susceptible to visual illusions while flying. Certainly, the dispassionate and robotic like automation of an AI self-driving car would not be prone to these human limitations, you might insist. Indeed, many proponents of AI self-driving cars keep saying that the wonderment of AI self-driving cars is that we don't need to worry about human drivers anymore that at times tend to drive while DUI or that get distracted while looking at their phones as they are behind the wheel.

First, let's clarify that a true self-driving car, considered a Level 5 self-driving car, is at a level at which the AI is supposed to be able to fully drive the car without any human intervention needed, and that the Level 5 consists of AI that can drive in whatever manner a human could drive a car. At the levels less than 5, the human is still considered the driver of the car, even if the automation or AI is there doing driving too. In that sense, the less than level 5 cars are still reliant upon humans, and so whenever the AI hands over the controls to the human driver, or whenever the human driver opts to take over the controls from the AI, we're now back in the realm of dealing with human susceptibility to sensory illusions.

But, I don't want to distract away from the very important point here, specifically that the AI is also susceptible to sensory illusions.

If you need to sit down for a moment, now that I've mentioned this key aspect, please do so. Yes, in spite of the talk about how perfect the AI is going to be, the reality is that the AI and the self-driving car will also be susceptible to sensory illusions. It is going to happen. It has most certainly already happened. It is a danger. It is a known danger. It is something that the auto makers and tech firms aren't necessarily talking about. It is something that needs to be of great concern by all, and we need to put in place as many protective measures about it was we can.

What kind of sensory illusion could an AI self-driving car be susceptible to? Lots.

There are numerous sensors on an AI self-driving car. There are cameras that capture still images and video images. There are radar devices and sonar devices. There might be LIDAR (light and radar) devices. Etc. Each of these sensory devices is not perfect. Each of these sensory devices can be faulty. Each of these sensory devices can work as expected, and not be in any error condition, and yet nonetheless provide sensory data that is ambiguous. The AI needs to take the sensory data and make some logical sense out of it.

Take a look at my framework about AI self-driving cars as additional background on the matter.

The sensors collect data about the real-world around the self-driving car. This data is then usually transformed into something more amenable for the AI to deal with. For example, the data of an image might be very large in terms of the number of pixels, along with some pixels being unspecified or being captured but considered unsure of whether they are on/off, and the raw data is so voluminous that the AI wouldn't be able to fully inspect it per se, and thus there might be a transformation and compression of the data that the sensor software undertakes. Thus, whatever originally was captured is not necessarily what the AI is about to try and interpret.

The transformed data is fed into the AI that's running the self-driving car. The AI needs to figure out whether there's a pedestrian in the image that was just captured. Similar to my earlier story about camping in the woods and whether you saw a bear or a human, the AI needs to try and guess from the image whether there's a pedestrian standing in the street up ahead or not. Maybe its just a cone in the street. Maybe it's a child. Maybe it's a pedestrian but they are actually further away than the image suggests. If the image is captured at nighttime, the darkness might make the shape of the pedestrian hard to fully distinguish. In short, the chances of a sensory illusion are quite substantial.

In fact, you might want to read my analysis of the Uber incident in Arizona, since it is possible that the self-driving car might have encountered a sensory illusion that led to it striking and killing the pedestrian that was walking a bicycle.

Is that a Tractor Up Ahead?

Another factor to keep in mind about the sensory illusion of an AI self-driving car involves the Machine Learning (ML) elements. Suppose we've used a Machine Learning approach such as an artificial neural network and trained the neural network on identifying cars based on images of cars. This is akin to having a neural network learn to identify cats in images, which the neural network might do so by perhaps identifying that cats seem to have a certain kind of ear shape and they have whiskers. Thus, when an image of something is fed into the neural network, and if the image has what looks to be cat ears and cat whiskers, the neural network would report that it has found a cat.

Suppose we fed images of the rear ends of cars into a Machine Learning element such as a neural network, doing so to allow that when the AI self-driving car takes a picture of a car up ahead, the neural network can take that rear end image of the car and try to figure out whether it is a car and what type of car it is.

We might have thousands upon thousands of images of the rear ends of Ford cars, BMW's, and so on. They all are used to train the neural network. It thusly eventually tunes until it is able to somewhat reliably detect that the image contains a rear end of a car in it.

Now, let's go back to the cat and let's say I fed an image of cat that had no whiskers and its ears were oddly shaped (unlike any normal cat). The neural network that had been trained on conventional cat images would be unlikely to identify that there's a cat in the image. In that same manner, if the AI self-driving car is driving along a dirt road, and a tractor is up ahead, the self-driving car upon inspecting the image of what the object is, might not be able to ascertain that it is a tractor. The rear end image of what a tractor looks like would be a lot different looking than the rear end image of a conventional car.

In this instance, I'm willing to include this example into the sensory illusion basket, even though we might all agree that the image of the tractor is let's say unobscured and fully visually detected. Technically it is not really a sensory illusion as we might normally consider a sensory illusion. I am allowing it to be considered as such to point out that what the AI is looking for and what it finds can be two different things. Here, the AI is looking for whether a car is ahead, and it finds something that does not seem to be a car, but we would likely agree it essentially is a car in that it is a mode of transportation that acts like car acts.

This scenario can play out in ways that are quite dangerous. The AI might assume that since its seemingly not a car ahead, maybe it can be ignored. Or, maybe it makes some other untoward assumption. Whatever action the AI decides to take regarding the self-driving car, there is now a heightened chance that any maneuvers might be poorly chosen ones. Some believe that perhaps the now famous case of the Tesla that slammed into the truck on a Florida highway in 2016, might have involved a sensory illusion issue. It was claimed that perhaps the side of the truck was perceived by the Autopilot automation as being the sky, and so the AI of the Tesla did not presumably think any object was up ahead.

One of the concerns about using Machine Learning techniques such as neural networks is that the neural network might "learn" aspects that don't really necessarily make logical sense to what we assume the neural network presumably has identified. For the cats example, suppose the neural network had found a pattern that suggested that cats all have brown fur (pretend that the images used for training purposes were of predominantly cats that had brown fur). In that case, the neural network might report that anytime a cat image is shown later on, unless it also has brown fur, the neural network indicates that the aspect of a cat being in the image is low or nonexistent.

There was a famous case of a ML system that examined images of military tanks and was shown pictures of Russian tanks and United States tanks. The neural network seemed to be able to differentiate them. Turns out that the Russian tank photos had lots of graininess while the US tanks did not, and the neural network patterned on the graininess of the images, and not on the actual distinctive features of a tank.

Returning to the Yanny versus Laurel debate, it's a fun topic and allows us all to enjoy kidding each other about what we hear and what we think we hear. It also though fortunately and interestingly brings up the importance of sensory illusions. Humans are susceptible to sensory illusions. AI self-driving cars are also susceptible to sensory illusions.

Let's not delude ourselves into thinking that AI self-driving cars are some kind of perfection. We need to develop the AI capabilities to be able to catch its own susceptibly to sensory illusions. Unlike the rather idle consequences of whether you hear Yanny or Laurel in that now ubiquitous audio clip, sensory illusions for AI self-driving cars can have decisive life-and-death consequences. It's a serious matter, and I assure you that's no illusion.

CHAPTER 11

FEDERATED

MACHINE LEARNING

AND

SELF-DRIVING CARS

Lance B. Eliot

CHAPTER 11

FEDERATED MACHINE LEARNING AND SELF-DRIVING CARS

Machine Learning (ML) is essential for the advent and further progress of AI self-driving cars. The nature of how Machine Learning is being undertaken today for AI self-driving cars will undoubtedly evolve and become more sophisticated over time. One crucial aspect for Machine Learning in the context of AI self-driving cars is whether or not to distribute out the Machine Learning aspects, and if so to what degree the ML should be distributed.

This aspect of distributing ML is often referred to as Federated Machine Learning (FML). You can think of the word "federated" in the same sense that it is used for the governmental arrangement of the United States. The United States is a collection of distributed States that are collectively part of an overarching federation. We are continually striving in the United States to ascertain what is the appropriate balance of States rights versus Federal, and there are ongoing debates about how much autonomy the States should have and versus how much control the federal government should have. This applies in the same manner to Federated Machine Learning, as will be further explored herein.

Allow me to first provide a quick story that perhaps helps illustrate the notion of federated learning.

I had done some high-tech consulting work for Snap-on Tools a number of years ago. You might be aware that Snap-on Tools is a famous brand of high-end tools that are often used by car mechanics in automotive repair and maintenance shops. Most car mechanics love their Snap-on Tools. They tend to be very passionate about how great the tools are, how durable they are, and so on. Indeed, many car mechanics are essentially fans of Snap-on Tools and enjoy wearing a Snap-on branded cap or shirt, and are proud to proclaim that they buy and use Snap-on Tools.

What is especially interesting about Snap-on Tools here is that they are primarily a franchise based business and make use of "dealers" to actually sell the tools. You've perhaps seen the Snap-on Tools trucks driving around town (they are very distinctive in appearance). The franchise network consists of over 4,000 such dealers. They drive around town, visiting the automotive repair and maintenance shops. During a visit, the dealer will open the back of the truck, pull down a ramp, and try to get the mechanics to take a break from work and come into the truck to see the tools that are being sold.

When I did my high-tech work at the firm, I actually went on several rides with some of the dealers. It was amazing to see the car mechanics get as excited to see the truck as they would if it were an ice cream truck or a food truck. The mechanics would usually rush out of the repair bays and relish coming to see the tools. They would salivate at the new tools and would often dream that someday they could buy a full set of Snap-on Tools. The enthusiasm for the brand was intense. You might liken this to people that are avid fans of Apple products and gush at the sight of a new iPad or iPhone.

I was putting in place an intranet that would allow the franchisees to readily communicate with each other. Up until then, they had no easy means to communicate with each other, beyond the use of email and some crude email distribution lists. By putting in place an intranet, it would enable all the franchisees to become more aware of what was taking at the firm and in the field. It would also aid franchisees in terms of assisting each other. They could be in their truck, driving around town, and when they came to a stop at a site visit, they could check the intranet to see if there were any useful announcements or other aspects posted there.

To try and showcase the value of the new intranet site, I went with one of the dealers just as we had rolled it out. He did his usual thing of driving to an auto repair facility, put down the ramp, and invited the mechanics to come on in. But, he did something else that I hadn't seen done before. He had taped a competitor's cap to the ramp. At first, this seemed odd to me, since I figured why in the world would he want the mechanics to see another brand name and be thinking about anything other than Snap-on. Well, there was a method to his madness.

One by one, the car mechanics came to the ramp, and as they walked up the ramp, each one stopped for a moment, took their boot and smashed down on that cap, moving their foot back and forth like they were trying to squash a bug. They delighted in doing this. I realized that the dealer had found an easy way to reinforce the devotion to Snap-on. As loyalists, they were able to showcase their support by walking all over a competitor brand. Every mechanic that walked up the ramp did so. It was also clever because some of them came into the back of the truck simply because they wanted to have their turn smashing down on the cap. They otherwise might not have come into the truck and instead stood outside and just bemoaned the fact that they couldn't afford another tool just then. By getting them into the truck, the allure of the tools would potentially get them into a buying mood and overlook the price.

I suggested to the dealer that he share this trick with the other franchisees. He was easily able to do so via the intranet. He posted his approach and right away got some comments that it seemed like a good idea. Within two days, a large portion of the franchisees had adopted this simple technique. The odds are that without the intranet to allow for convenient communication electronically, almost no one else would have known about the trick. Some might have discovered it on their own, and maybe a few might have learned about it via word-of-mouth, but otherwise it would not have become so widespread.

This is the potential power of federated learning.

When something is learned at an outside edge, it can be conveyed to the larger federation, and the federation can possibly propagate it out to the rest of the collective. There are learnings that can occur solely at the federation that are then shared with the edges. There are learnings at the edges that can be shared with the federation, and then possibly embraced throughout. This prevents an otherwise isolated learning from becoming "trapped" within an edge, and never seeing the light of day beyond its being used at that particular edge.

What does this have to do with AI self-driving cars?

At the Cybernetic Self-Driving Car Institute, we are in the midst of developing and advancing the use of Federated Machine Learning for AI self-driving cars.

Take a look at Figure 1 on the next page.

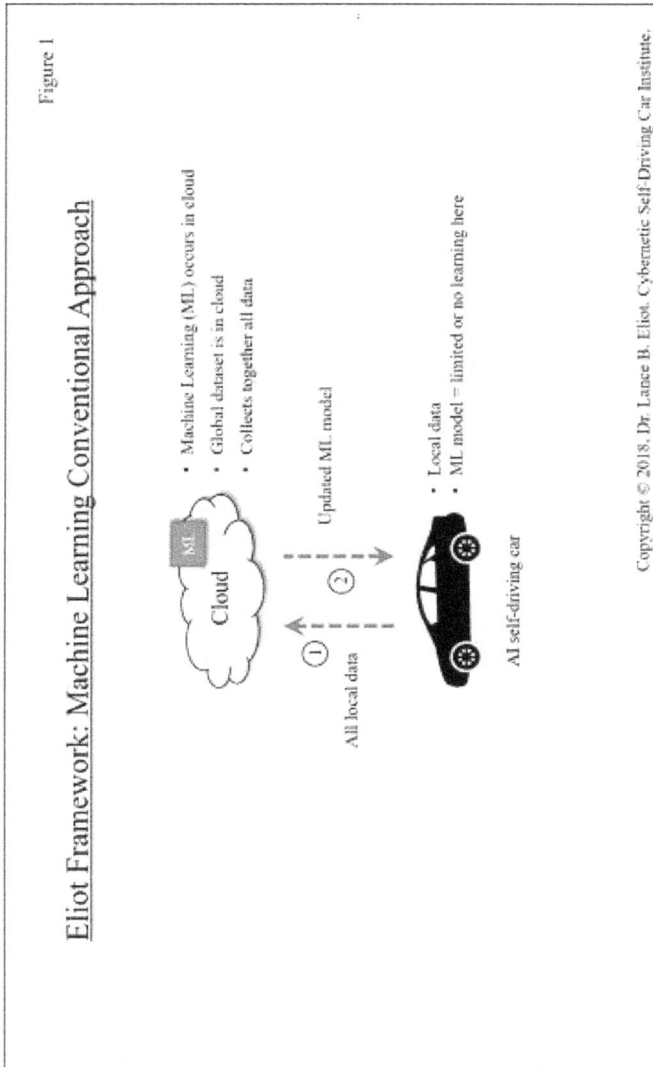

Figure 1

Eliot Framework: Machine Learning Conventional Approach

- Machine Learning (ML) occurs in cloud
- Global dataset is in cloud
- Collects together all data

Cloud

Updated ML model

All local data

AI self-driving car

- Local data
- ML model = limited or no learning here

Copyright © 2018, Dr. Lance B. Eliot, Cybernetic Self-Driving Car Institute.

As shown in Figure 1, the conventional approach right now to Machine Learning for most AI self-driving cars is that the Machine Learning happens in the cloud.

The AI self-driving car, we'll consider it an edge device, provides data that is uploaded to the cloud (typically this cloud would be setup by the auto maker or tech firm providing the AI capabilities for the self-driving car). The Machine Learning takes place in the cloud and then the resultant updated ML model is pushed down into the AI self-driving car. This happens via the OTA (Over The Air) capabilities of the AI self-driving car.

For this conventional approach, there really isn't much of a federation taking place per se. It is simply that each of the AI self-driving cars that are included in this collective are dutifully uploading their collected data, and the real action of using that data for ML purposes occurs in the cloud. Presumably, all of the AI self-driving cars then get the resultant updated ML model and are obligated to use it, once it has been pushed down into the AI self-driving car locally.

There have been some concerns raised that with essentially all of the data being uploaded from the AI self-driving car, there are perhaps privacy aspects that are being shared into the cloud that otherwise don't need to be (well, at least don't need to be uploaded for the purposes of the Machine Learning that is going to take place). Suppose the self-driving car is keeping track of how many times you've visited your local emergency room, because you have some ailment, does that kind of data really need to be shared with the cloud for purposes of doing Machine Learning on how to enhance the AI driving capability? Some would argue that there's a lot of data collected by the AI self-driving car that does not and should not be shared into the cloud.

Take a look at Figure 2 on the next page.

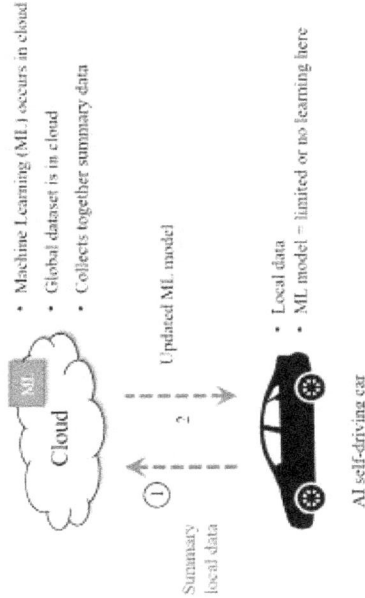

Figure 2

Eliot Framework: Machine Learning Conventional Approach

- Machine Learning (ML) occurs in cloud
- Global dataset is in cloud
- Collects together summary data

Cloud — ML

Updated ML model

Summary local data

- Local data
- ML model = limited or no learning here

AI self-driving car

If you believe in this concern for privacy, it could be the case that the data uploaded would only be summary data and also only data that's pertinent directly to the purposes of doing Machine Learning or for other intended and identified legitimate purposes.

Besides the privacy aspects, this also would substantially cut down on the transmission time of conveying the data and would presumably be less taxing on any electronic communications established for cloud connecting elements.

The next evolution of this ML would be to actually become more federated and work in some kind of collaborative mode with the edges and the cloud.

Take a look at Figure 3 on the next page.

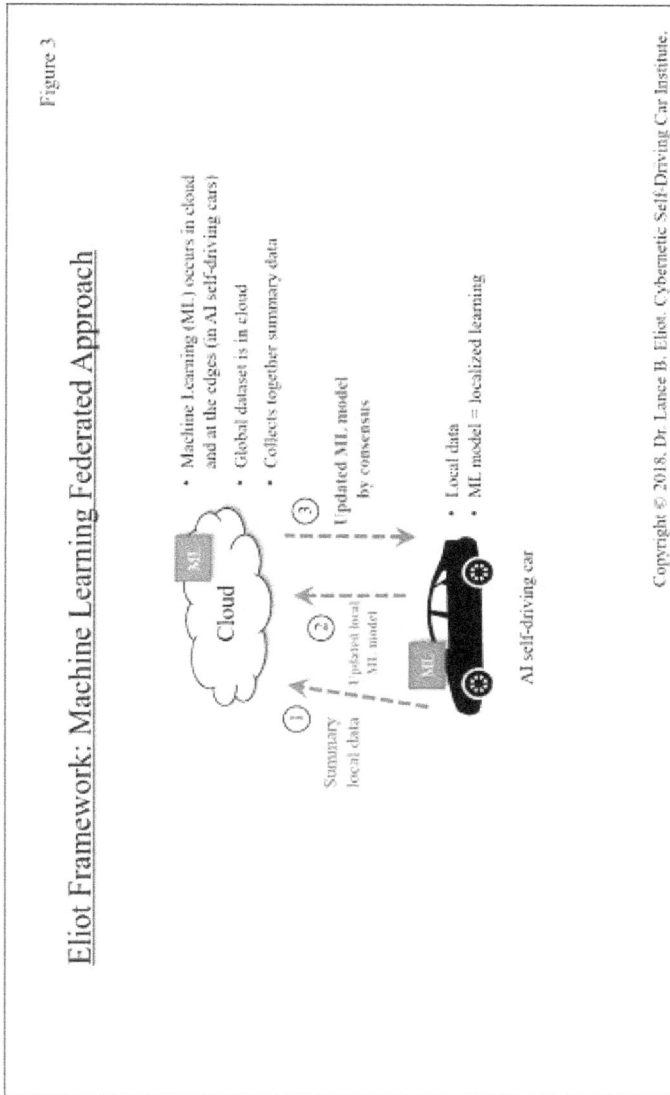

Figure 3

Eliot Framework: Machine Learning Federated Approach

Cloud
ML

- Machine Learning (ML) occurs in cloud and at the edges (in AI self-driving cars)
- Global dataset is in cloud
- Collects together summary data

① Summary local data

② Updated local ML model

③ Updated ML model by consensus

AI self-driving car
ML

- Local data
- ML model = localized learning

As shown, the Machine Learning in the AI self-driving car is also doing actual Machine Learning, and it then provides an updated ML up to the cloud.

The cloud then has to figure out what to do with this updated ML, along with having the summarized data from the AI self-driving car too. The cloud-based Machine Learning can potentially use the now-provided updated ML model from the AI self-driving car, further expand or refine it, and then ultimately push it back down to the AI self-driving car, which would replace the prior ML model with the new one. Alternatively, the push could be just the changes of a new version of the updated ML model that has been modified via efforts in the cloud.

It will be unlikely that there is only one AI self-driving car in the collective or federation. Instead, it is assumed that there will be lots of AI self-driving cars in a particular federation.

Take a look at Figure 4 on the next page.

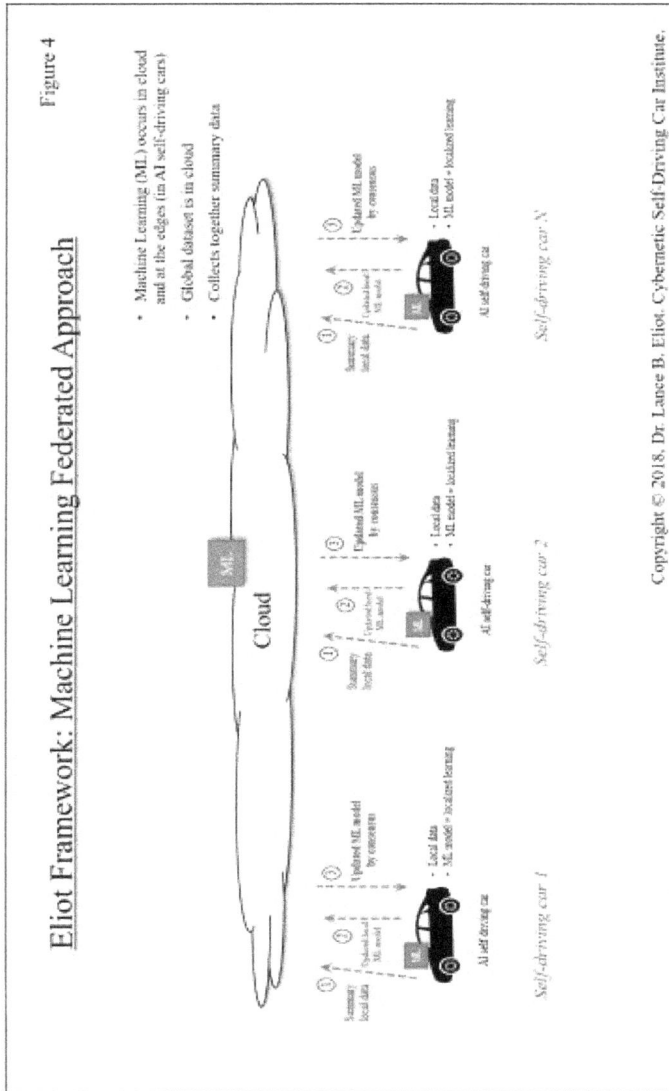

Figure 4

Eliot Framework: Machine Learning Federated Approach

- Machine Learning (ML) occurs in cloud and at the edges (in AI self-driving cars)
- Global dataset is in cloud
- Collects together summary data

Cloud

Self-driving car 1

Self-driving car 2

Self-driving car N

Copyright © 2018, Dr. Lance B. Eliot, Cybernetic Self-Driving Car Institute.

Here, you can see that there are some multiple number of AI self-driving cars, each of which has its own respective ML model, each of which is collecting its own local data, and each of which is providing up to the cloud it's local summarized data and its latest updated ML model. The cloud now has the "collective wisdom" from the set of AI self-driving cars.

You might call this the wisdom of the crowd approach, or more formerly it is referred to as a Federated Machine Learning Architecture.

Implementing this is a lot harder than it looks. You need to consider a wide array of aspects of how you want to architecture this. There is a myriad of trade-offs.

Let's start with some fundamentals. Will all of the AI self-driving cars in the federation be ending up with the same ML model, upon each of the refresh cycles? In other words, we could force all of the AI self-driving cars to have the same ML model, which would provide consistency across the AI self-driving cars and greatly simplify matters. On the other hand, this might also then negate some localized ML model aspects that would otherwise be helpful or maybe even crucial to a particular AI self-driving car or subset of the AI self-driving cars.

I know you could say that everyone gets everything, regardless of any localized aspects, and that the AI self-driving car in a localized context would then use that localized element but otherwise would not care about it. This though also suggests that we can actually isolate the localized difference and prevent it from being invoked when perhaps it would be best to not have it invoked. There is also the concern that the ML model becomes excessively large and unwieldly, and also might become a transmission hog if it is entirely being sent out with each refresh cycle.

This dovetails into a related question, namely how will the localized ML models be comingled into a single comprehensive ML model? Some say that there should be a consensus approach. If enough of the localized ML models seem to have reached a new state that offers value, it gets included. Meanwhile, if the localized ML model is the only one or less than a consensus worth of having something new, it does not get included.

Returning to my earlier story about Snap-on Tools, the adoption of the "smashing the cap" originated with one dealer. If, at first, the rule was to not allow an isolated instance to be propagated, it might not have gotten an opportunity for traction with the other dealers.

162

After a few dealers started to embrace it, there was a Yelp-like review scoring that caught the attention of the other dealers, and helped propel it into the lexicon of many of the dealers.

So, we need to figure out whether something new or interesting at a local ML model is worthwhile to possibly include into the federated ML model, or whether it should be extinguished, or whether it is retained but only for the contributing ML model or some subset of the AI self-driving cars. This is not easy to programmatically do with aplomb.

We next need to consider the security aspects. Suppose somehow a localized ML model gets infiltrated with something bad, such as an indication that once the AI self-driving car goes over 50 miles per hour and sees a green street sign it should then direct the self-driving car to crash into the nearest wall. Let's pretend this gets pushed up to the cloud. Let's further assume that the cloud realizes this is something new, but doesn't have any particular means to determine the reasonableness of it, and opts to then include it into the federated ML model. The federated model gets pushed out to the AI self-driving cars. Voila, this is like allowing a malware virus to readily be shared, doing so via the handy mechanism put in place for more valid purposes.

Thus, the security of what goes up, how it is encrypted, what comes down, and how the ML model and the data are stored in the AI self-driving car, in the cloud, and in transit, are all essential to this federated approach working properly. Ensuring tight security will be crucial. The advantage of having the OTA capability can be readily turned into a huge disadvantage is overtaken for evil purposes.

Another aspect involves conflicts among the ML models that are provided to the cloud. Let's say that one ML model indicates to never go faster than the speed limit, while another ML model has indicated that sometimes going faster than the speed limit is warranted such as an emergency to rush the occupants to a hospital. How again will the cloud be ascertaining what stays and what goes, and if model updates are in contradiction how to resolve the contradiction.

If you are further interested in this topic, you might find of interest that Google is doing some fascinating work on Federated Machine Learning as it relates to mobile devices such as smart phones. They have considered too aspects of optimization algorithms, such as the Stochastic Gradient Descent approach, and how it can be utilized in this Federated ML structure.

How much ML should be taking place in the AI self-driving car itself? Should it be getting data from the cloud that was collected from other AI self-driving cars, so it can do further ML? How much processing do we want to take place in an AI self-driving car and how much will that raise the cost and complexity of the AI self-driving car? These and many other questions are now being explored, and the future of AI self-driving cars as being part of a Federated Machine Learning approach is a necessity and still open matter of intense and vital focus.

CHAPTER 12
IRREPRODUCIBILITY
AND
SELF-DRIVING CARS

CHAPTER 12

IRREPRODUCIBILITY
AND SELF-DRIVING CARS

One of the most vital tenets of science consists of the principle of reproducibility.

When a scientific result is reported, how do we know that it is something generalizable and not just a happenstance one-time fluke or possibly even merely the outcome of a mistake made in performing the science? We can have a greater belief in the result if it is possible to reproduce the science. Reproducibility consists of several factors, including perhaps foremost that the effort when replicated will independently achieve either the same results or similar to.

If I claim to have invented a perpetual motion machine, and I demonstrate it in my lab, but doing so only for myself and perhaps some chosen lab assistants, I would want to tell the world of this amazing breakthrough. The rest of the world might be very elated to know that finally a perpetual motion machine has been created. Just think of how it could change the world. Others that have tried to devise a perpetual motion machine are excited too, but also a bit wary since they were unable to achieve the same grandiose outcome on their own. Naturally, they would want to know how this perpetual motion machine works, and they would want to try and reproduce it in their own labs.

I might then reproduce the effort in my own lab. I then come back to the world and say that yes, it really does work. Voila! But, should you believe that it really does work? Notice that the reproducibility was not independently carried out. Instead, the same researcher or developer merely claimed that they were able to reproduce the result. It could be very sincere and the researcher or developer is genuine in their belief that they reproduced the results. What the rest of the world doesn't yet know is whether it is really bona fide, or maybe an unfortunate mistake made repeatedly, or maybe an attempt at tricking the world into believing something that just isn't true.

Cold Facts About Cold Fusion

So that you don't think that this is an absurd notion of something like this happening, I'd like to remind you of the now classic case of desktop cold fusion.

For those of you that were around in the late 1980s or have studied the history of science, you might know that in 1989 there were two chemists at the University of Utah that claimed they had been able to generate cold fusion. The chemists, Stanley Pons and Martin Fleischmann, said that they had produced excess heat that had to be as a result of a nuclear process, a miniature nuclear reaction as it were. The equipment used was relatively inexpensive, easy to assemble, and pretty much what you could do in your own backyard (with a bit of effort, albeit). It became one of the biggest news stories of the year and was heralded as an incredible breakthrough.

Did anyone try to reproduce cold fusion? You betcha.

Scientists all around the world scampered to see if they could also achieve cold fusion. Some did so to help prove that the Utah scientists were right and that the discovery was valid. Others did so to disprove the Utah claims and to ensure that the world would not be falsely misled. Some did so just out of curiosity as to how it works and what else it might lead us to.

Unfortunately, much of the details about how the Utah scientists achieved their result was kept somewhat secretive and thus it made it difficult to try and reproduce. The number of reproducing efforts that said it could not be reproduced grew quickly, while the few that said they had been able to reproduce it were later chagrined to say they had to withdraw their results after further assessment. There were also later on the discovery of various sources of error in the original work, which tarnished it even more so. A real kill shot occurred when it was indicated that there weren't any detected nuclear reaction byproducts as a result of the alleged cold fusion (the byproducts should have appeared if the cold fusion worked as claimed).

Bottom-line: No one has yet to truly reproduce it per se, and so it now sits on the junk heap of what some consider unproven, unsound, infamous scientific work.

Maybe this cold fusion is an aberration and we shouldn't focus on one isolated incident involving scientific reproducibility. In that case, you might want to consider reading an article in the notable journal *Science* in 2015 that provided a research study done by independent scientists that were trying to reproduce the efforts of 100 of the most prominent studies in psychology. According to their independent work, they could only successfully reproduce about 39% of the studies. There have been various similar studies that have tried to reproduce other well-known and well-accepted scientific studies and been unable to do so entirely.

Please don't toss out the baby with the bath water on this topic. Anyone that reaches the conclusion that all science is fake news, well, you've got to revisit that kind of mentality since it doesn't make any sense here. There is lots and lots of science that is fully bona fide. Some of it has been reproduced, some it has not been, and for the portion that's not been reproduced you cannot conclude it is therefore invalid. You can only say that is hasn't been reproduced.

This too can be misleading because often times the same kind of study is not specifically reproduced, and instead there are other studies that build upon the results of the original study. Thus, you could assert that those extensions beyond the original study should presumably not have been valid if they were based on an original invalid core. If the extension studies are valid and being reproduced, you could assert that the original core was valid.

The potential unpleasant aspect of this involves circumstances where further down the road it is somehow shown that the original work was invalid, which can then call into question all of the other work that followed upon it. Fruit borne of the poisonous tree, as they say. Knocking down a house of cards, some might suggest.

Reproducibility Is Not So Alluring

Why isn't every science work being subjected to reproducibility?

That's an easy answer – there's not much incentive for doing reproducibility studies. As a scientist, you usually are measured by how much new science you create. Spending your time on simply reproducing someone else's work is not going to get you much mileage. If you showcase that the original work is valid, you aren't breaking new ground and instead merely adding to the belief that the original work was novel and sound. If you showcase that the original work was invalid, the odds are that you'll immediately be under attack by the original researchers and others that believe in the soundness of the original effort. Until others jump on your bandwagon of the invalid nature of the work, you are likely to be an outcast in the science community.

Situations like cold fusion are unusual in that any kind of incredible breakthrough is going to immediately draw attention by powerful forces in support and in opposition. There's nothing more disconcerting and motivating that if you've spent most of your scientific life trying to achieve a perpetual motion machine, and unsuccessful in doing so, and then someone claims they did so, you would put your entire energy into wanting to prove them wrong. How could they have done it, when you labored for forty years and could not do so? Their claims must be false. You'll gain notoriety for having shown it to be false, and also settled those that might look at you and otherwise say that you wasted those forty years looking in the wrong place to find that breakthrough.

Again, furthermore, anyone with such an investment in an area of inquiry is also going to want to know whether the breakthrough works or not generally anyway, so I don't want to overly emphasize the downside of things.

How does this reproducibility apply to AI self-driving cars?

At the Cybernetic Self-Driving Car Institute, we are urging the auto makers and tech firms to make available the AI self-driving car efforts they are undertaking, as best they can, and especially urge academic researchers and commercial research outfits to publish the details of their efforts, so that a form of reproducibility can be undertaken overall.

At this juncture, there is almost no reproducibility taking place in the AI self-driving car realm.

Work by most AI developers in self-driving cars is either considered proprietary and not revealed publicly, or those doing the work are so hectic that they have no time to "waste" on doing reproducibility and are instead trying to get these systems out the door, so to speak.

For the auto makers and tech firms, it is a tough call as to whether to showcase the innards of their AI self-driving car systems. Each of those firms are spending millions upon millions of dollars to develop AI self-driving car capabilities. Why should they just hand it over to anyone that wants it? The proprietary nature of what they are doing creates enormous value for their firm and they rightfully should be able to seek payoff of the tremendous investments they are making.

I say this because there are some AI developers that are decrying the secretive nature of these firms. But, who can blame these firms that have made these huge investments? There is a race to see who gets to the moon first, so to speak, and these private entities are betting the farm on getting there. It makes no logical sense for them to simply handout their secret sauce. Intellectual Property (IP) in the AI self-driving car field is king.

On the other hand, we have the argument that if each of them develops their respective AI self-driving car capabilities in isolation of each other, and if there isn't some means for others to independently verify what they are doing, how can we be satisfied that what these self-driving cars will do on our public roadways is safe?

There are some that argue that for these AI self-driving cars to be allowed onto the public roadways the auto makers and tech firms should be forced to open the kimono. If they want our roads to test on, they need to share what they've got. Otherwise, got off our roads. The counter-argument is that we might not ever see true self-driving cars if the auto makers and tech firms need to use only private test areas or even government sponsored proving grounds. The amount of self-driving car experience that a self-driving car can gain on a providing ground is considered a tiny fraction of what it can gain as experience in the real-world of everyday driving.

It's the classic madcap desire for new innovation that is being weighed against the costs of getting there.

Academic Research Reproducibility

Even the academic AI researchers are being accused of some less than forthcoming reveals about their work. For many of the studies on Machine Learning (ML) and the use of neural networks, there are often claims of incredible breakthroughs on being able to do vision recognition or speech recognition or whatever, and yet the actual neural network is not made available for anyone else to try and independently verify. Are we to take at face value whatever is reported by the researchers?

Sometimes, the researchers will point to the fact that their work was published in a peer review journal. These researchers then assert that by doing so it shows that their work must be valid. Not so fast. For most peer reviews, the peer is not trying to actually reproduce the results of the study they are reviewing. Instead, the peer reviewers are supposed to examine the study and try to ascertain whether it seems sound and valid as based on whatever happens to be presented by the researchers. Peer reviewers that are experts in the field have a body of knowledge to help judge whether the nature of the study and the outcome seem to be plausible. But, they are almost never doing an actual reproducibility prior to agreeing whether to accept the research for publication or not.

There are studies that suggest peer reviews aren't necessarily as rigorous as one might assume or hope for. There are often inherent biases by peer reviewers. Suppose the prevailing wisdom is that the world is flat. Suppose you are an in expert in the flat earth field. You get a research study that further supports that the world is flat. The odds of you rejecting it are low. Suppose you get a study that says the earth is round. You might reject it due to the assertion that the author is obviously not aware of the acceptance that the earth is flat. And so it goes. I am not saying all peer reviewers should be painted with this same brush, and I'd like to emphasize that many peer reviewers do an amazing job and try to remain impartial as they do so.

Studies of the body of "peer reviews" also have tended to suggest that often times the peer reviewer is either not well versed in analyzing the statistics found in scientific studies, or does not take the time to assess the statistics used. This means that a scientific study might get a free pass on how they did the statistics, which also could mean that the results are not statistically significant and thus could be considered questionable or even possibly invalid. Indeed, there is a program called Statcheck, produced out of Tilburg University, which purports to analyze the statistics used in science studies and then be able to find potential errors in how the statistics were used. This has brought controversy, including that sometimes the results of the assessment have been posted online and yet not allowed the original authors to rebut what is being posted.

For AI self-driving cars, there have been some initial efforts to go the open source route with the AI system component, including posting the source code and the neural network models being used. Generally, these are not the full effort and often more of a scaled-down version of what is taking place. Again, you can hardly blame the auto makers and tech firms from not wanting to go the open source route for their AI self-driving cars.

Analogy to Pharmaceuticals

If a pharmaceutical company wants to bring a new life-saving drug to the marketplace, they need to undergo a rigorous, costly, time consuming, and secrets-revealing process. Some say that we should treat AI self-driving cars in the same manner. You can certainly see the comparison in the sense that a new drug can mean life-or-death for people. Likewise, an AI self-driving car can mean life-or-death for people. Would you be willing to take a drug that was untested and had not gone through a rigorous process?

Of course, the drug safety question is perhaps even less so than the AI self-driving car safety question. If you as an individual take an unproven drug and it kills you, that's one life lost. If an AI self-driving car is unsafe, it can kill the human occupants, human pedestrians, and humans in other cars. If a drug is distributed into society and we discover its bad, generally it can be pulled from shelves and word will be spread to avoid taking the drug. If a self-driving car is unsafe, you can't so easily pull it from use, since it means that people that are depending upon it for mobility are now somewhat immobilized.

The counter argument by AI developers and auto firms is that via OTA (Over The Air) updates, you can readily make an unsafe AI self-driving car into presumably a safe one. This use of an electronic beamed set of changes is unlike what could be done about a bad drug. There's no means to do anything about a bad drug other than get rid of it and issue a new drug. An unsafe AI self-driving car can be remotely transformed into a safe one, presumably.

This strident belief of the OTA as a solver of unsafe issues for AI self-driving car is not quite the savior it might seem to be.

The aspect that a particular AI self-driving car is unsafe needs to be first discovered, and it could be that there is a lurking bug in the AI of self-driving cars that causes those cars to generate numerous deaths before it can be figured out what the problem is. Even once the problem is figured out, a bug fix needs to be crafted in order to do the OTA with it. The AI self-driving cars need to receive and enable the

OTA fix. It could be that the fix causes other unsafe conditions to arise. Suppose that there are many such bugs, and there are only being discovered one at a time, and thus this unsafe AI self-driving car is repeatedly undergoing this cycle of killing people, then an effort to find the bug, an effort to fix it, an effort to possibly fix the fix, and so on. For an unsafe AI self-driving car, this can just keep repeating over and over again.

And, I've already repeatedly called for more transparency in the AI self-driving car field.

How could this transparency work? There are mechanisms being formulated to allow for collecting together a body of code and its artifacts, and then making them available for others to try using. For example, the Jupyter Notebook is a promising open-source web application for binding together code, visualizations, text, models, machine learning, and other artifacts. This provides a means to embody the work. Places such as GitHub are great as a venue for placing such work.

Some say that the AI self-driving car field should be treated like the airline industry and there should be various rules and regulations about safety in the same manner as are tracked for airplanes. For example, the Flight Data eXchange (FDX) is a special de-identified aggregated collection of data that captures safety related events about airplanes. It contains data from over 100 airports and allows for doing analysis of safety aspects. The AI self-driving car field is still in its infancy and not anywhere near the maturity of the airline industry, but it would seem instructive to learn from the airlines and try to proactively get ourselves ready for safety aspects. Maybe we need to have a AI Self-Driving Car Data Exchange (AISDCX).

You might say that the states that are allowing AI self-driving cars are requiring the auto makers and tech firms to report safety related data, thus, we've already solved this aspect. Not quite.

Overall, the nature of the reporting has been considered relatively insufficient and not at all like the FDX kind of approach. Some say that if we burden the auto maker and tech firms with overly onerous reporting and regulatory requirements, it will kill the goose that laid the golden egg. Other says it is something that needs to come with the turf.

In the academic realm of AI self-driving cars research, we are gradually seeing more acceptance of the pre-publishing or pre-print approach to posting research prior to it being subjected to the rigors (and possibly delays) of peer reviews.

Some applaud this as a means to try and get new science into the hands of the community as quickly as possible. This is especially important in an area like AI self-driving cars wherein advances in machine learning are moving very quickly and it would handy to leverage the work of others right away.

We are also seeing more of a call for computer science researchers to post their code and models, rather than just referring to their work in their write-ups. Doing so will aid the reproducibility factor. It will also aid those that are trying to stand taller on the shoulders of others, and aid those neophytes coming up the ranks that want to learn from others.

Irreproducibility is an important limiter in the progress of science and engineering.

For those of you that are new to the subject, you might find of interest the rather large body of research about the value of reproducibility as essential to innovation (easily Googled); while for those of you that have been steeped in the reproducibility topic for years, there are a smattering of rather hilarious looks at irreproducibility, including the long standing Journal of Irreproducible Results (JIR) which publishes made-up scientific studies (it's funny stuff!), and also some of the even more staid and bona fide journals will occasionally publish intentional irreproducible stories as an April Fool's prank.

Though I enjoy a good laugh from time-to-time, irreproducibility is a real problem facing the AI self-driving car field and we're going to have to tackle it, one way or another.

CHAPTER 13

IN-CAR DELIVERIES
AND SELF-DRIVING CARS

CHAPTER 13

IN-CAR DELIVERIES
AND SELF-DRIVING CARS

Quick, tell me what's in the trunk of your car.

Like most people, you probably have quite a hodgepodge of items in your car's trunk. When my children were young, I had sports gear that they needed after classes finished, changes of clothes for them, various school supplies like extra sets of writing paper and marker pens, and whatever else made sense to transport for them.

When I was in college, I used to do a lot of road trips to visit friends at other universities and so I kept a sleeping bag, basketball, volleyball, bowling ball, skateboard, foldable bike, and lots of other personal items in the trunk of my car. Indeed, there was pretty much no available space in my trunk, and when I went to the airport to pickup someone that was coming to town, I'd have to insist that their baggage went with them into the car rather than fitting into the trunk (well, to my credit, I did have cords in the trunk that I could use to tie down luggage to the roof of the car).

A trunk is a pretty handy thing.

Sometimes you put old stuff in it. Sometimes you put new stuff in it. Sometimes it is jam packed to the gills, other times it as empty as can be.

You might recall the famous comedian George Carlin that did a hilarious bit about the fact that we all have stuff. Our houses are full of our stuff. Our cars are full of our stuff. When we go on vacation, we take some of our stuff with us. At the hotel while on vacation we leave some of our stuff while we go on an excursion, and then make sure to come back to our stuff. Maybe it seems materialistic, but I'd say that we like our stuff and we want to make sure our stuff is safe and sound.

Speaking of which, you might remember that when the idea of ordering stuff on the web was first coming to the attention of society, there were these wild predictions that we'd all need to get some kind of specialized box put at the front of our homes and that could be used to deliver stuff to us. You can't fit that coffee maker that was ordered online into your conventional mailbox, plus legally your mailbox is only supposed to be used for certain stipulated purposes, and so the thought was that we would all purchase large sized, separately positioned lockboxes for our homes.

These lockable containers would be on our front porch, or easier still for the delivery aspects would be to have the containers out near the street. A delivery could then happen by a delivery truck that would just pull up to the container, and by entering a code would open it and then place the shipment into the open space. The deliverer would then close the container and somehow lock it, either by its own automatic means or by the deliverer taking some kind of locking action. I remember predictions that there would be millions upon millions of these home-based shipping container boxes contraptions that would be sold within just a few years' time.

These containers had some potential downsides.

Would they look ugly and turn our pleasant neighborhoods into a sea of commerce looking boxes, was one such concern. Another was how to secure the container itself. In other words, if you have this container thing sitting on your front porch, and if thieves realize that some nifty items are being placed into it, might the thieves just follow along after the delivery truck and haul away your entire container in their own truck? Presumably, they would take it to some dinghy warehouse and break open the container to then take your goods and opt to resell it on e-bay or elsewhere.

Delivery At the Front Door Has Its Drawbacks

When the containers approach fizzled, we all began to get used to just having delivery trucks place boxes of ordered items at our front door. But, this has the same thievery aspects associated with it. Furthermore, if you were to leave the delivered items at your door for a day or two, home burglars might reason that you are away and so they might come and break into your house looking for even better loot. Taking the thievery aspects out of it, having boxes of delicate items sitting at your front door also makes those boxes susceptible to the elements. The most hardened of cardboard boxes can still have troubles when dosed in several inches of rain.

We've identified so far then that we want to get stuff delivered to us, and we aren't keen on buying a special container to sit outside our house, and we don't want to have stuff just dropped at our doors. What next? As you likely know, the next approach involved allowing a delivery to take place into your home.

You've perhaps seen the experiment that Amazon has been doing of having deliveries occur into your home. You put a special electronic lock on your door, the delivery person uses an electronic code to open the door, they put the delivery into your house, you can see them on your in-house web cam, and they then are supposed to politely leave your house, not disturbing anything, and the door should be locked after they head-out.

Some scratch their head about the idea of letting a delivery person into their home when the home is otherwise unoccupied. For many, entry into their home is a sacred act. They don't like the idea of a stranger, even one that is authorized, coming into their homes. But, there are some people that think it's a great idea. They love the idea that they can get a delivery into the safe environs of their home. What excitement to come home after a long day's work and see that apparently Santa has dropped off some items you ordered. It's like

Christmas each and every day!

The odds are though that the percentage of people that are going to use the in-home delivery method is relatively low. If that's not a popular method, and if leaving stuff at your door is not desired, and if getting a container for outside your home is not fruitful, you might be wondering what other possibility exists.

Answer: Use the trunk of your car.

Think about this idea for a moment. Your trunk is a relatively safe place. You already keep all sorts of important items in there. Suppose you parked your car at the front of your home, and allowed a delivery to occur that gets placed into the trunk of your car. If the delivery could occur with some allowed remote unlocking of your car, you wouldn't need to be nearby to receive the delivery. You could be in your house, sound asleep, and meanwhile a delivery comes up to your parked car, opens the trunk, drops in a package, closes the trunk, makes sure the car is locked, and voila, you now can get the item out of your trunk whenever you want. Cool.

This car-as-container idea is handy too since presumably your car goes where you go.

In other words, you drive to work and park your car nearby to your office. You are in the office for let's say eight hours. During that time, you could potentially have something delivered to the trunk of your car. Indeed, you could have multiple deliveries possibly made to your trunk. When you finally get done at the office, you drive your car home and all of the neat new stuff you got delivered is right there in your trunk.

Amazon recently made quite a news splash by announcing that they are going to provide in-car delivery services. There are other delivery firms that have been considering the same idea, but you've got to hand it to Amazon to be the first big-time proponent and announcer of this new, shipping delivery approach. It's another reason for people to shop at Amazon. If it becomes popular, you can bet that other online firms and even brick-and-mortar firms will soon be offering the same kind of service (they will have to, for competitive purposes).

How does the in-car delivery work?
Take a look at Figure 1.

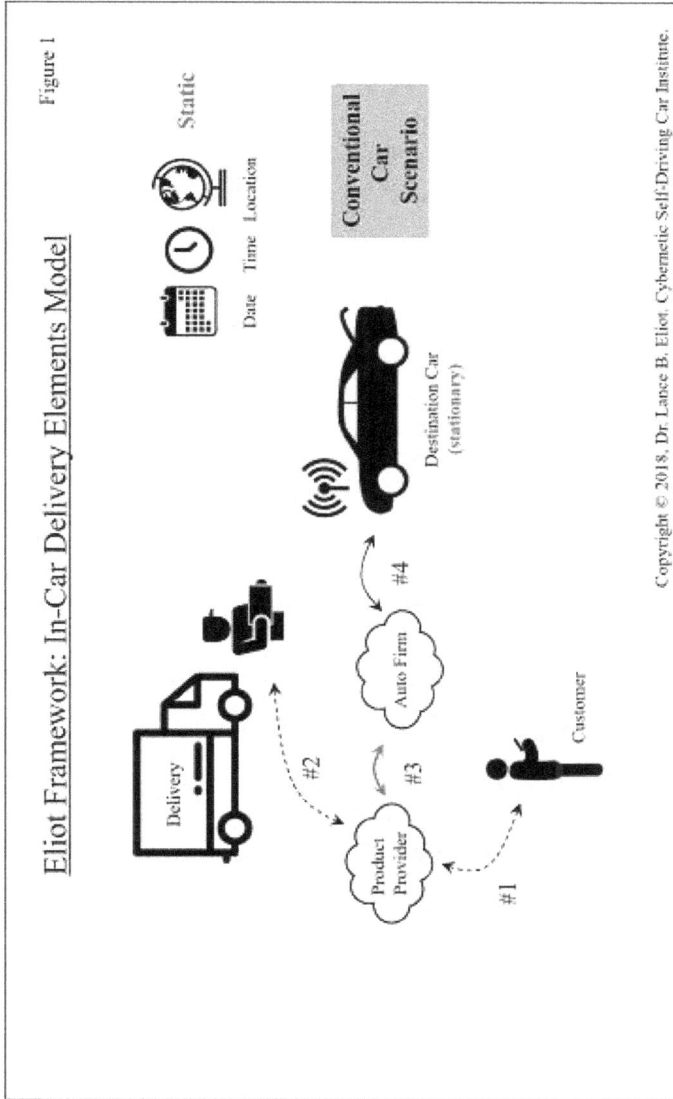

Basically, the customer first downloads a special app, which for Amazon it's the Amazon Key app and only usable by Amazon Prime members currently. The customer links their "product provider" cloud service (in this example, Amazon) with their connected car service such as OnStar. For certain models of cars (Amazon is doing this with certain newer models of GM and Volvo autos), the customer will be able to specify that they want to have a delivery done to their car.

The delivery will be scheduled in a particular time window, of which Amazon is saying four hours for now and we'll see if that gets tightened up, and when the deliverer arrives at the vehicle, the car will get unlocked by communicating to the connected car service, the deliverer than puts the item into the trunk, and either directly re-locks the car or the connected car services does so (with the approach now being that the deliverer is supposed to lock the car, but either way the connected service does so after some amount of time once the delivery has occurred, doing so as a fail-safe method).

On the one hand, it's pretty simple.

I'd also say that there's opportunity for things to go awry. With any such new service, there will likely be some hiccups and shake-out as it gets perfected.

Your car is supposed to be parked in a public place that is readily accessed by the deliverer. I am sure there will be cases of people parking their car in a restricted area and not realizing that the deliverer can't get to their car, such as in a gated community or in a private parking structure that doesn't otherwise allow public entry. I am guessing there will be instances of people having their car parked in a bad area, and once the deliverer shows up, either some bad people try to steal from the deliverer, or once the deliverer drives away that those bad people might try to break into the car or maybe even steal the car.

There are likely going to be circumstances when the deliverer finds the right car, and upon opening the trunk discovers there is insufficient space to put the package. This could be because the trunk is already full. Or, it could be that someone ordered a refrigerator and had it delivered to the trunk of their car (don't underestimate dolts!).

This also raises the question about the nature of trunks themselves. Some vehicles don't have a true trunk and instead have an area in the body of the car that has a shield to try and cover the so-called trunk area. If the deliverer puts your boxes there, it is possible that any passerby will see the boxes that you have, which could either be an intrusion of your privacy, or again be a lure to break into your car.

Further issues involve the status of the boxed item once it is in your car. Suppose the deliverer properly deposited it, but then the sun beats down on your car the whole day, and when you come out to get your ordered item, it's a melted goo.

There are lots of other "oddball" (some say "edge") cases too.

Suppose you have left Aunt Betty in the car, and the delivery takes place – should the deliverer interact with Aunt Betty or just open the trunk and put in the package? Suppose you've left a dog in your car (which you shouldn't do), will that impede the deliverer? Should the deliverer actually report you to the authorities if they find the dog stuck inside your car? Suppose the deliverer opens the trunk and finds fifty pounds of illegal drugs, what should they do?

In-Car Delivery Considerations

Here's some more.

Suppose you've illegally parked your car. Should the deliverer still try to make the deposit.

By the way, where is the deliverer going to park the delivery truck while delivering to your car? Will the delivery car sit double-parked in the street while trying to make the deposit? We could be encouraging dangerous driving behavior by this method of shipment and delivery.

Suppose your car is parked in a manner that getting the trunk is nearly impossible (it's butted up against a wall or some other car). What then?

Will the deliverer find the right car? In theory, they should, since they are presumably going to have a description of the car, plus the license plate info, plus the connected car system which will unlock the car (in spite of those aspects, I suppose it's still possible that deliveries to the wrong car might happen). As a twist, suppose that I offer to you, my office colleague, that you can use my car to get your delivery today. In this case, will whatever security provisions allow this, or would the fact that you are the customer, and you are specifying a car you don't own or control, would that be permitted or not accounted for.

Some wonder about the safety factor of this too. We are usually all watching for suspicious people near cars. If someone looks like they are wearing a delivery uniform, will we now accept that they might be standing at your car and trying to do a delivery (or might be masquerading as such to break into it)?

The 4-hour time window is another interesting facet. Suppose you park your car, intending to leave it for the four hours. But, an emergency occurs, and you opt to use your car. Will the deliverer get notified, which this also brings up how often your car will be beaming its location to the deliverer? Privacy experts worry that people will be giving up their routine of where they park their car, which could be used for maybe advertising purposes (the provider starts flooding you with ads for that coffee shop across the street from where you seem to be normally parking your car), or worse.

We'll also need to deal with things like breakage.

Suppose the delivery truck accidentally brushes against your car while trying to make the delivery. And/or, suppose the human deliverer accidentally bangs your car while making the delivery. Suppose the deliverer shoves the package into the trunk, and you had your precious collection of fine china in the trunk, which now is crushed by the package. Or, maybe you look in your trunk and find the package there, but the fine china is now missing (did the delivery person take it?).

In short, there will be lots and lots of exceptions, and we'll have to see how much it tarnishes the whole notion of the approach.

If the first people that use the service get big news coverage that something went awry, it might force the firms wanting to do this to go into a retreat. The retreat might involve being more selective about who and what gets delivered, or maybe that only certain areas get deliveries, etc. Hopefully, whatever maladies occur, it won't end-up killing the goose. Generally, this idea of using the trunk of our cars as a shipment destination seems appealing and I would anticipate it will become quite popular.

What does this have to do with AI self-driving cars?

At the Cybernetic Self-Driving Car Institute, we have already been anticipating the in-car delivery prospects and have been doing various development aspects accordingly (yes, yet another edge problem to be solved for AI self-driving cars).

Let's consider why an AI self-driving car as an in-car delivery is any different than a conventional car.

Before we jump into the whiz bang aspects, keep in mind that the AI self-driving car could act like it is a conventional car, meaning that it simply sits and waits for a delivery to occur. In this mode, it does not exercise any of its AI self-driving capabilities. I want to mention this so that it is clear cut that the AI self-driving car can still be used as a conventional car for the in-car delivery purposes.

But, with its "super powers" the AI self-driving car can do more.

First, a conventional car is going to be stationary at a fixed location for the time window that the shipping delivery is supposed to occur.

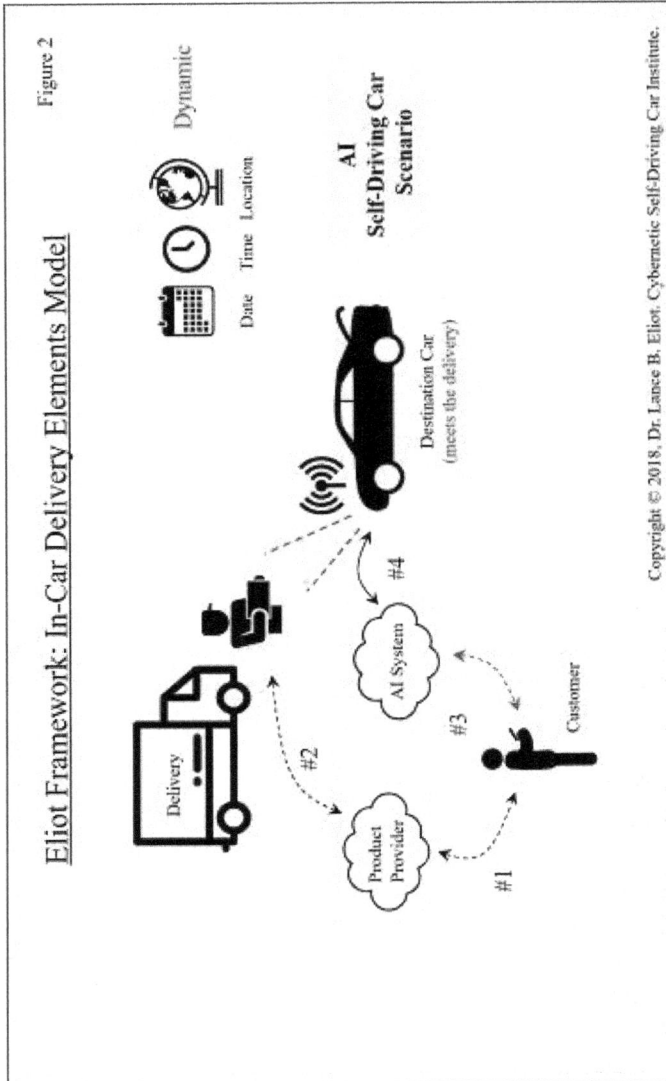

Eliot Framework: In-Car Delivery Elements Model

Figure 2

AI Self-Driving Car Scenario

Dynamic

Date Time Location

Destination Car (meets the delivery)

AI System #4

#3

Customer

Delivery #2

Product Provider #1

As shown in Figure 2, the AI self-driving car does not necessarily need to be sitting around. Keep in mind that the AI self-driving car essentially has a "driver" sitting in it, ready for use, at all times. As such, the AI self-driving car could opt to go toward the delivery vehicle and meet it at some mutually agreed and convenient place.

Suppose you parked your car at your company private parking structure. While you are at the office, your AI self-driving car leaves the parking structure, and via electronic communication with the delivery cloud, it finds a place that is good for both the deliverer and the AI self-driving car. The AI self-driving car comes up to the delivery truck, and they both park momentarily for the human deliverer to then put the package into the AI self-driving car. The AI self-driving car then drives back to your company parking lot, parks, and waits for you to next make use of the self-driving car.

I realize that some of you will say that why even have a human deliverer, and though yes at some future time we might have some more robotic way to do this, in the nearer term it seems reasonable to assume that there is still a human delivery person involved. Whether the human deliverer is also driving the delivery vehicle will be a second question, in the sense that the delivery vehicle might also be an AI self-driving vehicle. This would certainly change the dynamics too of who could be a human deliverer, since they might not need to have a driver's license anymore for the purposes of driving the delivery vehicle.

You might be wondering why does the AI self-driving car go to the deliverer, and instead why not go to say the warehouse where the goods are stored. The AI self-driving car could certainly go to a warehouse and pick-up a package, but let's assume for the moment that it is a longer drive, and that instead there are these delivery vehicles acting as a distributed method of having goods floating around. It would likely make more sense for the AI self-driving car to meet with the floater, rather than having to make a trip to a fixed location warehouse.

Another subtle difference of the conventional car versus the AI self-driving car involves what happens after the delivery occurs. Suppose you want to get a shipped item to your children at their school, and so your AI self-driving car goes to their school (after having met with the deliverer), your children get the item out of the trunk, and then your self-driving car comes to your workplace to wait there until you are done for the day.

The point being that the AI self-driving car can act in a mobility fashion, doing so both prior to the delivery and at post-delivery.

This capability also changes the static nature of the 4-hour time window for delivery into instead providing a flexible and dynamic time alignment. Nearly at any time, the coordination and synchronizing of the deliverer and the AI self-driving car can occur, thus, there might be an agreed overall window of general time and location, which could be firmed-up and allow then that the AI self-driving car doesn't have to be pinned down to a static location and time per se.

Let's also consider some other ramifications of using the features of an AI self-driving car.

You could potentially turn your AI self-driving car into your own delivery vehicle. Suppose that a delivery of several items for people on your block are available from the true delivery vehicle, and in coordination with your neighbors, your AI self-driving car picks those up from the true delivery vehicle, and then your AI self-driving car goes to each house in your neighborhood to allow your neighbors to get their respective item out of your trunk.

Yet another difference involves how your conventional car interacts with the product provider. It could be that rather than using a connected car service per se, you are able to control your AI self-driving car via the auto maker cloud. You then might be directly coordinating with the product provider. No need to necessarily have the connected car service acting as a middleman.

More differences exist.

When the deliverer comes to a conventional car, the conventional car might not have any ability to signal to the deliverer that they have found the correct car. For an AI self-driving car, it could become conspicuous by turning the headlights on-off, or honking the horn, etc. This might help the deliverer to more readily find your car and also increase the odds of getting the package to the correct car.

The AI self-driving car can also use its sensors to the advantage of undertaking the delivery. With the cameras on the self-driving car, it could visually record the activity of the deliverer, which might be handy in case of any disputes. Your AI self-driving car could also live stream video to you, and thus you could watch as the deliverer makes the delivery (similar to the idea of having a web-cam in your house for watching in-home deliveries).

Do we want to use our cars as a place to store our stuff?

That's the fundamental question. Since we already do keep stuff in our cars, it makes sense to allow for shipments to be delivered to our cars. While we perceive our homes as sacred ground, I'd guess that most people perceive their trunk as less hallowed. The convenience of having something delivered to your car is quite appealing. Adding into the mix the capabilities of AI self-driving cars would seem to further up the ante. We'll need to see whether people accept the idea of in-car delivery for conventional cars, and if so, I'd suggest that the added benefits of doing so with an AI self-driving car might just knock off their socks. By the way, those are socks that were delivered to the trunk of your car.

Lance B. Eliot

APPENDIX

APPENDIX A
TEACHING WITH THIS MATERIAL

The material in this book can be readily used either as a supplemental to other content for a class, or it can also be used as a core set of textbook material for a specialized class. Classes where this material is most likely used include any classes at the college or university level that want to augment the class by offering thought provoking and educational essays about AI and self-driving cars.

In particular, here are some aspects for class use:

o   Computer Science. Studying AI, autonomous vehicles, etc.

o   Business. Exploring technology and it adoption for business.

o   Sociology. Sociological views on the adoption and advancement of technology.

Specialized classes at the undergraduate and graduate level can also make use of this material.

For each chapter, consider whether you think the chapter provides material relevant to your course topic. There is plenty of opportunity to get the students thinking about the topic and force them to decide whether they agree or disagree with the points offered and positions taken. I would also encourage you to have the students do additional research beyond the chapter material presented (I provide next some suggested assignments they can do).

RESEARCH ASSIGNMENTS ON THESE TOPICS

Your students can find background material on these topics, doing so in various business and technical publications. I list below the top ranked AI related journals. For business publications, I would suggest the usual culprits such as the Harvard Business Review, Forbes, Fortune, WSJ, and the like.

Here are some suggestions of homework or projects that you could assign to students:

a) <u>Assignment for foundational AI research topic</u>: Research and prepare a paper and a presentation on a specific aspect of Deep AI, Machine Learning, ANN, etc. The paper should cite at least 3 reputable sources. Compare and contrast to what has been stated in this book.

b) <u>Assignment for the Self-Driving Car topic</u>: Research and prepare a paper and Self-Driving Cars. Cite at least 3 reputable sources and analyze the characterizations. Compare and contrast to what has been stated in this book.

c) <u>Assignment for a Business topic</u>: Research and prepare a paper and a presentation on businesses and advanced technology. What is hot, and what is not? Cite at least 3 reputable sources. Compare and contrast to the depictions in this book.

d) <u>Assignment to do a Startup:</u> Have the students prepare a paper about how they might startup a business in this realm. They must submit a sound Business Plan for the startup. They could also be asked to present their Business Plan and so should also have a presentation deck to coincide with it.

You can certainly adjust the aforementioned assignments to fit to your particular needs and the class structure. You'll notice that I ask for 3 reputable cited sources for the paper writing based assignments. I usually steer students toward "reputable" publications, since otherwise they will cite some oddball source that has no credentials other than that they happened to write something and post it onto the Internet. You can define "reputable" in whatever way you prefer, for example some faculty think Wikipedia is not reputable while others believe it is reputable and allow students to cite it.

The reason that I usually ask for at least 3 citations is that if the student only does one or two citations they usually settle on whatever they happened to find the fastest. By requiring three citations, it usually seems to force them to look around, explore, and end-up probably finding five or more, and then whittling it down to 3 that they will actually use.

I have not specified the length of their papers, and leave that to you to tell the students what you prefer. For each of those assignments, you could end-up with a short one to two pager, or you could do a dissertation length paper. Base the length on whatever best fits for your class, and the credit amount of the assignment within the context of the other grading metrics you'll be using for the class.

I mention in the assignments that they are to do a paper and prepare a presentation. I usually try to get students to present their work. This is a good practice for what they will do in the business world. Most of the time, they will be required to prepare an analysis and present it. If you don't have the class time or inclination to have the students present, then you can of course cut out the aspect of them putting together a presentation.

If you want to point students toward highly ranked journals in AI, here's a list of the top journals as reported by *various citation counts sources* (this list changes year to year):

o   Communications of the ACM

o   Artificial Intelligence

o   Cognitive Science

o   IEEE Transactions on Pattern Analysis and Machine Intelligence

o   Foundations and Trends in Machine Learning

o   Journal of Memory and Language

o   Cognitive Psychology

o   Neural Networks

o   IEEE Transactions on Neural Networks and Learning Systems

o   IEEE Intelligent Systems

o   Knowledge-based Systems

GUIDE TO USING THE CHAPTERS

For each of the chapters, I provide next some various ways to use the chapter material. You can assign the tasks as individual homework assignments, or the tasks can be used with team projects for the class. You can easily layout a series of assignments, such as indicating that the students are to do item "a" below for say Chapter 1, then "b" for the next chapter of the book, and so on.

a) What is the main point of the chapter and describe in your own words the significance of the topic,

b) Identify at least two aspects in the chapter that you agree with, and support your concurrence by providing at least one other outside researched item as support; make sure to explain your basis for disagreeing with the aspects,

c) Identify at least two aspects in the chapter that you disagree with, and support your disagreement by providing at least one other outside researched item as support; make sure to explain your basis for disagreeing with the aspects,

d) Find an aspect that was not covered in the chapter, doing so by conducting outside research, and then explain how that aspect ties into the chapter and what significance it brings to the topic,

e) Interview a specialist in industry about the topic of the chapter, collect from them their thoughts and opinions, and readdress the chapter by citing your source and how they compared and contrasted to the material,

f) Interview a relevant academic professor or researcher in a college or university about the topic of the chapter, collect from them their thoughts and opinions, and readdress the chapter by citing your source and how they compared and contrasted to the material,

g) Try to update a chapter by finding out the latest on the topic, and ascertain whether the issue or topic has now been solved or whether it is still being addressed, explain what you come up with.

The above are all ways in which you can get the students of your class involved in considering the material of a given chapter. You could mix things up by having one of those above assignments per each week, covering the chapters over the course of the semester or quarter.

As a reminder, here are the chapters of the book and you can select whichever chapters you find most valued for your particular class:

Lance B. Eliot

Companion Book By This Author

Advances in AI and Autonomous Vehicles: Cybernetic Self-Driving Cars

Practical Advances in Artificial Intelligence (AI) and Machine Learning
by
Dr. Lance B. Eliot, MBA, PhD

This title is available via Amazon and other book sellers

Companion Book By This Author

Self-Driving Cars: "The Mother of All AI Projects"

by Dr. Lance B. Eliot, MBA, PhD

This title is available via Amazon and other book sellers

Companion Book By This Author

Innovation and Thought Leadership on Self-Driving Driverless Cars

by Dr. Lance B. Eliot, MBA, PhD

Chapter Title

This title is available via Amazon and other book sellers

<u>Companion Book By This Author</u>

New Advances in AI Autonomous Driverless Cars Self-Driving Cars

by Dr. Lance B. Eliot, MBA, PhD

<u>Chapter Title</u>

1 Eliot Framework for AI Self-Driving Cars

2 Self-Driving Cars Learning from Self-Driving Cars

3 Imitation as Deep Learning for Self-Driving Cars

4 Assessing Federal Regulations for Self-Driving Cars

5 Bandwagon Effect for Self-Driving Cars

6 AI Backdoor Security Holes for Self-Driving Cars

7 Debiasing of AI for Self-Driving Cars

8 Algorithmic Transparency for Self-Driving Cars

9 Motorcycle Disentanglement for Self-Driving Cars

10 Graceful Degradation Handling of Self-Driving Cars

11 AI for Home Garage Parking of Self-Driving Cars

12 Motivational AI Irrationality for Self-Driving Cars

13 Curiosity as Cognition for Self-Driving Cars

14 Automotive Recalls of Self-Driving Cars

15 Internationalizing AI for Self-Driving Cars

16 Sleeping as AI Mechanism for Self-Driving Cars

17 Car Insurance Scams and Self-Driving Cars

18 U-Turn Traversal AI for Self-Driving Cars

19 Software Neglect for Self-Driving Cars

This title is available via Amazon and other book sellers

Lance B. Eliot

Companion Book By This Author

Introduction to
Driverless Self-Driving Cars

by Dr. Lance B. Eliot, MBA, PhD

Chapter Title

This title is available via Amazon and other book sellers

Lance B. Eliot

Companion Book By This Author

Autonomous Vehicle Driverless Self-Driving Cars and Artificial Intelligence

by Dr. Lance B. Eliot, MBA, PhD

<u>Chapter Title</u>

This title is available via Amazon and other book sellers

Lance B. Eliot

Companion Book By This Author

Transformative Artificial Intelligence Driverless Self-Driving Cars

by Dr. Lance B. Eliot, MBA, PhD

This title is available via Amazon and other book sellers

Disruptive Artificial Intelligence
and Driverless Self-Driving Cars

by Dr. Lance B. Eliot, MBA, PhD

This title is available via Amazon and other book sellers

Lance B. Eliot

ABOUT THE AUTHOR

Dr. Lance B. Eliot, MBA, PhD is the CEO of Techbruim, Inc. and Executive Director of the Cybernetic Self-Driving Car Institute, and has over twenty years of industry experience including serving as a corporate officer in a billion dollar firm and was a partner in a major executive services firm. He is also a serial entrepreneur having founded, ran, and sold several high-tech related businesses. He previously hosted the popular radio show *Technotrends* that was also available on American Airlines flights via their in-flight audio program. Author or co-author of a dozen books and over 400 articles, he has made appearances on CNN, and has been a frequent speaker at industry conferences.

A former professor at the University of Southern California (USC), he founded and led an innovative research lab on Artificial Intelligence in Business. Known as the "AI Insider" his writings on AI advances and trends has been widely read and cited. He also previously served on the faculty of the University of California Los Angeles (UCLA), and was a visiting professor at other major universities. He was elected to the International Board of the Society for Information Management (SIM), a prestigious association of over 3,000 high-tech executives worldwide.

He has performed extensive community service, including serving as Senior Science Adviser to the Vice Chair of the Congressional Committee on Science & Technology. He has served on the Board of the OC Science & Engineering Fair (OCSEF), where he is also has been a Grand Sweepstakes judge, and likewise served as a judge for the Intel International SEF (ISEF). He served as the Vice Chair of the Association for Computing Machinery (ACM) Chapter, a prestigious association of computer scientists. Dr. Eliot has been a shark tank judge for the USC Mark Stevens Center for Innovation on start-up pitch competitions, and served as a mentor for several incubators and accelerators in Silicon Valley and Silicon Beach. He served on several Boards and Committees at USC, including having served on the Marshall Alumni Association (MAA) Board in Southern California.

Dr. Eliot holds a PhD from USC, MBA, and Bachelor's in Computer Science, and earned the CDP, CCP, CSP, CDE, and CISA certifications. Born and raised in Southern California, and having traveled and lived internationally, he enjoys scuba diving, surfing, and sailing.

ADDENDUM

State-of-the-Art
AI Driverless Self-Driving Cars

*Practical Advances in Artificial Intelligence (AI)
and Machine Learning*

By
Dr. Lance B. Eliot, MBA, PhD

———

For supplemental materials of this book, visit:

www.ai-selfdriving-cars.guru

For special orders of this book, contact:

LBE Press Publishing

Email: LBE.Press.Publishing@gmail.com

www.ingramcontent.com/pod-product-compliance
Lightning Source LLC
Chambersburg PA
CBHW021556210326
41599CB00010B/463